Collapse of I-35W Highway Bridge
Minneapolis, Minnesota
August 1, 2007

ACCIDENT REPORT

NTSB/HAR-08/03
PB2008-916203

National
Transportation
Safety Board

NTSB/HAR-08/03
PB2008-916203
Notation 7975C
November 14, 2008

Highway Accident Report

Collapse of I-35W Highway Bridge
Minneapolis, Minnesota
August 1, 2007

**National
Transportation
Safety Board**

490 L'Enfant Plaza, S.W.
Washington, D.C. 20594

National Transportation Safety Board. 2008. *Collapse of I-35W Highway Bridge, Minneapolis, Minnesota, August 1, 2007.* Highway Accident Report NTSB/HAR-08/03. Washington, DC.

Abstract: About 6:05 p.m. central daylight time on Wednesday, August 1, 2007, the eight-lane, 1,907-foot-long I-35W highway bridge over the Mississippi River in Minneapolis, Minnesota, experienced a catastrophic failure in the main span of the deck truss. As a result, 1,000 feet of the deck truss collapsed, with about 456 feet of the main span falling 108 feet into the 15-foot-deep river. A total of 111 vehicles were on the portion of the bridge that collapsed. Of these, 17 were recovered from the water. As a result of the bridge collapse, 13 people died, and 145 people were injured. On the day of the collapse, roadway work was underway on the I-35W bridge, and four of the eight travel lanes (two outside lanes northbound and two inside lanes southbound) were closed to traffic. In the early afternoon, construction equipment and construction aggregates (sand and gravel for making concrete) were delivered and positioned in the two closed inside southbound lanes. The equipment and aggregates, which were being staged for a concrete pour of the southbound lanes that was to begin about 7:00 p.m., were positioned toward the south end of the center section of the deck truss portion of the bridge and were in place by about 2:30 p.m. About 6:05 p.m., a motion-activated surveillance video camera at the Lower St. Anthony Falls Lock and Dam, just west of the I-35W bridge, recorded a portion of the collapse sequence. The video showed the bridge center span separating from the rest of the bridge and falling into the river.

Major safety issues identified in this investigation include insufficient bridge design firm quality control procedures for designing bridges, and insufficient Federal and State procedures for reviewing and approving bridge design plans and calculations; lack of guidance for bridge owners with regard to the placement of construction loads on bridges during repair or maintenance activities; exclusion of gusset plates in bridge load rating guidance; lack of inspection guidance for conditions of gusset plate distortion; and inadequate use of technologies for accurately assessing the condition of gusset plates on deck truss bridges. As a result of this accident investigation, the Safety Board makes recommendations to the Federal Highway Administration (FHWA) and the American Association of State Highway and Transportation Officials. One safety recommendation resulting from this investigation was issued to the FHWA in January 2008.

Contents

GLOSSARY OF BRIDGE-RELATED TERMS AS USED IN THIS REPORT

This glossary also defines the <u>underlined</u> terms within definitions.

Abutment: A retaining wall that supports the ends of a bridge.

American Association of State Highway and Transportation Officials (AASHTO): A nonprofit association whose voting membership consists of representatives of the highway and transportation departments of every State, Puerto Rico, and the District of Columbia. AASHTO guides and <u>specifications</u> are used to describe loading requirements for highway bridges. The organization was formed in 1914 and, until 1973, was known as the American Association of State Highway Officials, or AASHO.

Approach span: The portion of a bridge that carries traffic from the land to the main parts of the bridge.

Bearing: A device located between the bridge structure and a supporting <u>pier</u> or <u>abutment</u>.

Box member: A hollow, rectangular, four-sided structural <u>member</u> "built up" from pieces of steel joined to form a "box." On the I-35W bridge, the upper and lower surfaces of these members were referred to as *cover plates*; the east and west sides were referred to as *side plates*. Transverse metal plates (<u>diaphragms</u>) were welded at intervals inside the box members to add rigidity.

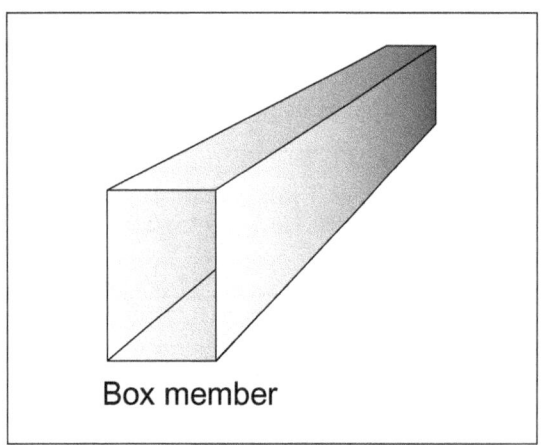

Box member

Bridge load rating: The live-load-carrying capacity of a bridge, determined by review of bridge plans and field inspection data. This rating is used to determine whether specific legal or overweight vehicles can safely cross the structure, whether the structure needs to be <u>load posted</u>, and the level of load posting required.

Cantilever: A structural member that projects beyond a supporting column or wall and is supported at only one end.

Compression: A force that pushes or presses toward the center of an object or from the ends toward the middle of a structural member.

Compression member: A truss member that is subjected to compressive (compression) forces. In the I-35W bridge, some structural members were always under compression; some were always under tension; and some, depending on the live load, reversed, changing from tension to compression or vice versa.

Condition ratings: According to the National Bridge Inspection Standards, condition ratings are used to describe an existing bridge compared with its condition when new. The ratings are based on materials, physical condition of the deck, superstructure, and substructure. General condition ratings range from 0 (failed condition) to 9 (excellent). Based on the bridge's condition, a *status* is assigned. The status is used to determine eligibility for Federal bridge replacement and rehabilitation funding. Current Federal Highway Administration status ratings are *Not Deficient*, *Structurally Deficient*, and *Functionally Obsolete*.

Culvert: A drain, pipe, or channel that allows water to pass under a road, railroad, or embankment.

Dead load: The static load imposed by the weight of materials that make up the bridge structure itself.

Deck: The roadway portion of a bridge, including shoulders. Most bridge decks are constructed as reinforced concrete slabs.

Deck truss bridge: A truss bridge with the truss underneath the roadway, supporting traffic traveling along the top of the main structure. In a through truss, traffic travels through the superstructure, which is cross-braced above and below traffic.

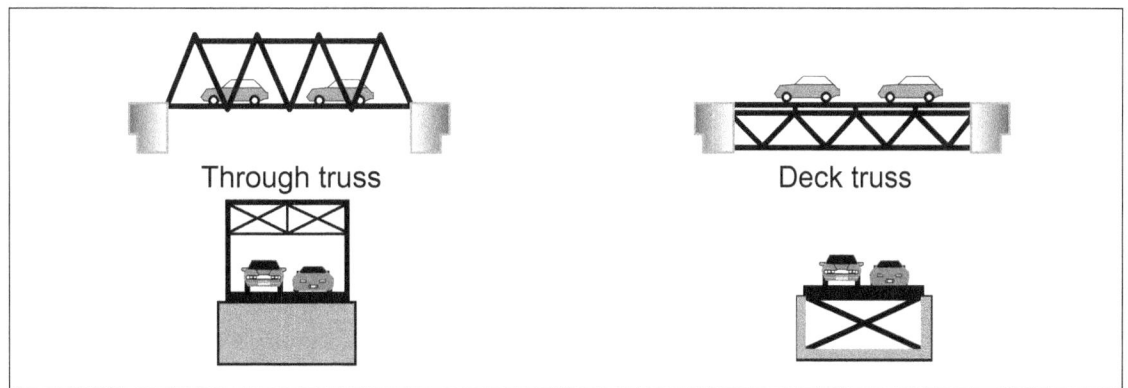

Through truss   Deck truss

Diagonal: A structural member connecting the upper and lower chords on the diagonal (as opposed to the vertical). See gusset plate.

Diaphragm: Bracing that spans between the main beams or girders of a bridge and assists in the distribution of loads. On the I-35W bridge, the box members also contained internal diaphragms.

Expansion joint: A meeting point between two parts of a structure that is designed to allow for independent movement of the parts due to thermal expansion while protecting the parts from damage. Expansion joints are commonly visible on a bridge deck as a hinged or movable connection perpendicular to the roadway.

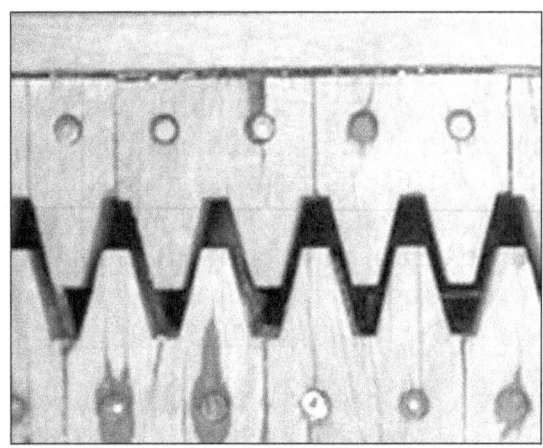

Typical expansion joint

Fatigue: In metal, a brittle cracking mechanism caused by repetitive loading over time.

Finite element analysis: An analysis of a structure based on a computer model of its material or design. A finite element model describes a virtual assembly of simplified structural elements used to approximate a complex structure. The behavior of the complex structure is then calculated by combining the actions of the interconnected simpler elements.

Floor truss: A welded truss perpendicular to the main trusses, used to support the deck.

Fracture-critical member: A steel member within a non-load-path-redundant structure, the failure of which would cause a partial or total collapse of the structure.

Functionally obsolete: A bridge status assigned by the Federal Highway Administration under the National Bridge Inspection Standards. A *Functionally Obsolete* bridge is one that was built to standards that are not used today. These bridges are not considered inherently unsafe, but they may have lane widths, shoulder widths, or vertical clearances that are inadequate for current traffic.

Girder: A horizontal structural member supporting vertical loads by bending. Larger girders are typically made of multiple metal plates that are welded or riveted together.

Gusset plate: A metal plate used to unite multiple structural members of a truss.

H member: A structural steel member with two flat flanges separated by a horizontal steel plate (<u>web</u>) to form an "H," made as either a rolled or welded shape.

Legal load: The maximum load for each vehicle configuration permitted by law by the State in which a State highway bridge is located. In the United States, the current weight limit for the Interstate system as set by the Federal Highway Administration is 80,000 pounds gross vehicle weight and 20,000 pounds for an axle.

I-35W bridge main truss node

Live load: Operational or temporary loads such as vehicular traffic, impact, wind, water, or earthquake.

Load: Force applied either from the weight of the structure itself (<u>dead load</u>) or from traffic, temporary loads, wind, or earthquake (<u>live load</u>).

Load posted: Any bridge or structure restricted to carrying loads less than the legal load limit. The <u>National Bridge Inspection Standards</u> require the load posting of any bridge that is not capable of safely carrying a <u>legal load</u>.

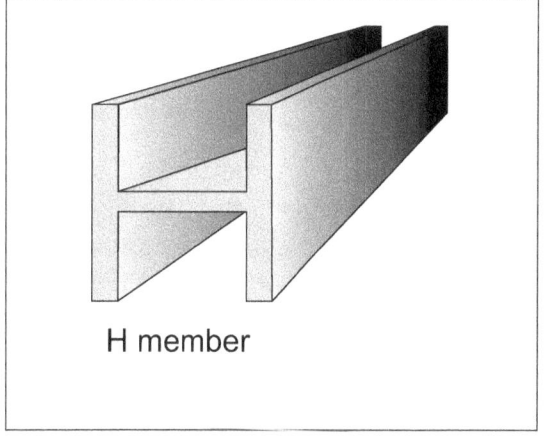

H member

Lower chord: The bottom horizontal, or almost horizontal, member of a truss. The lower chord extends the length of the deck truss but consists of shorter chord members spliced together at <u>nodes</u>.

Member: An individual angle, beam, plate, or built-up piece intended to become an integral part of an assembled frame or structure. The major structural elements of the truss (chords, diagonals, and verticals) are called members.

National Bridge Inspection Standards: Federal standards first established in 1971 to set national requirements for bridge inspection frequency, inspector qualifications, report formats, and inspection and rating procedures. The legislative authority for the standards is found at 23 *Code of Federal Regulations* Part 650.

Node: On the I-35W bridge, a connecting point where the upper or lower chords were joined to vertical and diagonal members with gusset plates. The bridge had 56 nodes on each of its two main trusses, for a total of 112 nodes. See the node illustration at gusset plate.

Nondestructive evaluation: Also referred to as nondestructive testing or nondestructive inspection, this evaluation does not damage the test object. Technologies for nondestructive evaluation include x-ray and ultrasound, which may be used to detect such defects as cracking and corrosion.

Non-load-path-redundant: The condition where fracture of an individual structural element (a fracture-critical element) could lead to a partial or total collapse of the entire bridge. A bridge that is non-load-path-redundant is not inherently unsafe, but it does lack redundancy in the design of its support structure. Such bridges are sometimes referred to as fracture critical. The I-35W bridge was of a non-load-path-redundant design.

Pier: A vertical structure that supports the ends of a multispan superstructure at a location between abutments.

Post-tensioning: A method of stressing concrete using steel rods or cables that are stretched after the concrete has hardened. This stretching of the rods or cables puts the concrete in compression, with the compressive stresses designed to counteract the tensile (tension) forces on the concrete once it is under load.

Rivet: A metal fastener, used in construction primarily before 1970, made with a rounded preformed head at one end and installed hot into a predrilled or punched hole. The other end was hammered (upset) into a similarly shaped head, thereby clamping the adjoining parts together.

Rocker bearing: A bridge support bearing that accommodates thermal expansion and contraction of the superstructure through a rocking action.

Roller bearing: A bridge bearing comprising a single roller or a group of rollers housed so as to permit longitudinal thermal expansion or contraction of a structure.

Section loss: A loss of metal, usually resulting from corrosion, that reduces the thickness of a steel bridge component.

Shear: A force that causes parts of a material to slide past one another in opposite directions.

Snooper: An inspection bucket or platform at the end of a long articulating boom (usually mounted to a truck) that provides access to the undersides of bridges.

Span: The horizontal space between two supports of a structure. A *simple span* rests on two supports, one at each end, the stresses on which do not affect the stresses in the adjoining spans. A *continuous span* comprises a series of consecutive spans (over three or more supports) that are continuously or rigidly connected (without joints) so that bending moment may be transmitted from one span to the adjacent ones.

A snooper being used to inspect the I-35W bridge

Specifications: A document that explains material and construction requirements of the bridge structure.

Splice plate: A plate that joins two chord members of a truss or that is used to extend the length of a <u>member</u>. On the I-35W bridge, the chord member joints used four splice plates—top, bottom, east, and west.

Stiffener: A structural steel shape, such as an angle, that is attached to a flat plate such as a gusset plate or the web of a member to add compression strength.

Stringer: A beam aligned with the length of a <u>span</u> that supports the <u>deck</u>.

Structurally deficient: A bridge status assigned by the Federal Highway Administration under the <u>National Bridge Inspection Standards</u>. A bridge is classified *Structurally Deficient* if it has a general condition rating for the deck, superstructure, substructure, or culvert of 4 (poor condition) or less.

A rating of *Structurally Deficient* does not indicate that the bridge is unsafe but that it typically requires significant maintenance and repair to remain in service and eventual rehabilitation or replacement to address deficiencies.

Substructure: The bridge structure that supports the <u>superstructure</u> and transfers loads from it to the ground or bedrock. The main components are <u>abutments</u>, <u>piers</u>, footings, and pilings.

Superstructure: The bridge structure that receives and supports traffic loads and, in turn, transfers those loads to the <u>substructure</u>. It includes the bridge deck, structural members, parapets, handrails, sidewalk, lighting, and drainage features.

Tension: A force that stretches or pulls on a material.

Tension member: Any member of a truss that is subjected to tensile (<u>tension</u>) forces. In the I-35W <u>truss bridge</u>, some structural members were always under tension; some were always under <u>compression</u>; and some, depending on the <u>live load</u>, reversed, changing from tension to compression or vice versa.

Truss bridge: A bridge typically composed of straight structural elements connected to form triangles. In large structures, the ends of the members are connected with <u>gusset plates</u>. Geometry ensures that the members are primarily loaded in direct <u>tension</u> or <u>compression</u>. The I-35W bridge structure was in the form of a Warren truss with <u>verticals</u>, which has alternating tension and compression <u>diagonals</u>.

Upper chord: The top horizontal, or almost horizontal, member of a truss. The upper chord extends the length of the deck truss, but it is made up of shorter chord members joined at <u>nodes</u>.

Vertical: The vertical member connecting the <u>upper</u> and <u>lower chords</u> at like-numbered <u>nodes</u>.

Wearing surface: The topmost layer of material applied on a roadway to receive traffic loads and to resist the resulting disintegrating action, also known as wearing course.

Web: The vertical portion of an I-beam or <u>girder</u>.

Yield stress: The stress above which permanent (plastic) deformation occurs.

EXECUTIVE SUMMARY

About 6:05 p.m. central daylight time on Wednesday, August 1, 2007, the eight-lane, 1,907-foot-long I-35W highway bridge over the Mississippi River in Minneapolis, Minnesota, experienced a catastrophic failure in the main span of the deck truss. As a result, 1,000 feet of the deck truss collapsed, with about 456 feet of the main span falling 108 feet into the 15-foot-deep river. A total of 111 vehicles were on the portion of the bridge that collapsed. Of these, 17 were recovered from the water. As a result of the bridge collapse, 13 people died, and 145 people were injured.

On the day of the collapse, roadway work was underway on the I-35W bridge, and four of the eight travel lanes (two outside lanes northbound and two inside lanes southbound) were closed to traffic. In the early afternoon, construction equipment and construction aggregates (sand and gravel for making concrete) were delivered and positioned in the two closed inside southbound lanes. The equipment and aggregates, which were being staged for a concrete pour of the southbound lanes that was to begin about 7:00 p.m., were positioned toward the south end of the center section of the deck truss portion of the bridge and were in place by about 2:30 p.m.

About 6:05 p.m., a motion-activated surveillance video camera at the Lower St. Anthony Falls Lock and Dam, just west of the I-35W bridge, recorded a portion of the collapse sequence. The video showed the bridge center span separating from the rest of the bridge and falling into the river.

The National Transportation Safety Board determines that the probable cause of the collapse of the I-35W bridge in Minneapolis, Minnesota, was the inadequate load capacity, due to a design error by Sverdrup & Parcel and Associates, Inc., of the gusset plates at the U10 nodes, which failed under a combination of (1) substantial increases in the weight of the bridge, which resulted from previous bridge modifications, and (2) the traffic and concentrated construction loads on the bridge on the day of the collapse. Contributing to the design error was the failure of Sverdrup & Parcel's quality control procedures to ensure that the appropriate main truss gusset plate calculations were performed for the I-35W bridge and the inadequate design review by Federal and State transportation officials. Contributing to the accident was the generally accepted practice among Federal and State transportation officials of giving inadequate attention to gusset plates during inspections for conditions of distortion, such as bowing, and of excluding gusset plates in load rating analyses.

Before determining that the collapse of the I-35W bridge initiated with failure of the gusset plates at the U10 nodes, the Safety Board considered a number of potential explanations. The following factors were considered, but excluded, as being causal to the collapse: corrosion damage in gusset plates at the L11 nodes, fracture of a floor truss, preexisting cracking, temperature effects, and pier movement.

The following safety issues were identified in this investigation:

- Insufficient bridge design firm quality control procedures for designing bridges, and insufficient Federal and State procedures for reviewing and approving bridge design plans and calculations.

- Lack of guidance for bridge owners with regard to the placement of construction loads on bridges during repair or maintenance activities.

- Exclusion of gusset plates in bridge load rating guidance.

- Lack of inspection guidance for conditions of gusset plate distortion.

- Inadequate use of technologies for accurately assessing the condition of gusset plates on deck truss bridges.

As a result of this accident investigation, the Safety Board makes recommendations to the Federal Highway Administration and the American Association of State Highway and Transportation Officials. One safety recommendation resulting from this investigation was issued to the Federal Highway Administration in January 2008.

FACTUAL INFORMATION

Accident Synopsis

About 6:05 p.m. central daylight time on Wednesday, August 1, 2007, the eight-lane, 1,907-foot-long I-35W highway bridge over the Mississippi River in Minneapolis, Minnesota, experienced a catastrophic failure in the main span of the deck truss. As a result, 1,000 feet of the deck truss collapsed, with about 456 feet of the main span falling 108 feet into the 15-foot-deep river. (See figure 1.) A total of 111 vehicles were on the portion of the bridge that collapsed. Of these, 17 were recovered from the water. As a result of the bridge collapse, 13 people died, and 145 people were injured.

The Accident

On the day of the collapse, roadway work was underway on the I-35W bridge, and four of the eight travel lanes (two outside lanes northbound and two inside lanes southbound) were closed to traffic. In the early afternoon, construction equipment and construction aggregates (sand and gravel for making concrete) were delivered and positioned in the two closed inside southbound lanes. The equipment and aggregates, which were being staged for a concrete pour of the southbound lanes that was to begin about 7:00 p.m., were positioned toward the south end of the center section of the deck truss portion of the bridge and were in place by about 2:30 p.m.

About 6:05 p.m., a motion-activated surveillance video camera at the Lower St. Anthony Falls Lock and Dam, just west of the I-35W bridge, recorded a portion of the collapse sequence. The video showed the bridge center span separating from the rest of the bridge and falling into the river.

A total of 111 vehicles were documented as being on the bridge when it collapsed.[1] Of these, 25 were either construction vehicles or vehicles belonging to construction workers. One of the nonconstruction vehicles was a school bus carrying 63 students and the driver. After the collapse, 17 vehicles were found in the river or on a submerged portion of the bridge deck.

---

[1] Trailers and nonmotorized construction-related equipment are not included in this total.

Figure 1. (Top) Aerial view (looking northeast) of I-35W bridge (arrow) about 2 hours 15 minutes before collapse. This photograph was taken by a passenger in a commercial airliner departing Minneapolis/St. Paul International Airport. (Bottom) I-35W bridge after collapse.

Emergency Response

Initial Response

About 6:05 p.m., Minnesota State Patrol dispatchers were notified of the accident by cell phone through the 911 system. After verifying the collapse using the Minnesota Department of Transportation (Mn/DOT) freeway camera system, dispatchers contacted the Minneapolis 911 dispatch, which is a combined emergency dispatch center for the Minneapolis fire and police departments. The first call from Minneapolis 911 dispatch went out at 6:07 p.m. At 6:08 p.m., Minneapolis 911 dispatch made a distress call over the interstate radio system requesting that all available emergency assistance providers respond to the I-35W bridge.

Some of the first to become involved in the rescue effort were citizens who were in or near the area when the collapse occurred. The Minneapolis police captain responsible for the on-scene investigation estimated that 100 citizens assisted in the total rescue effort. These people included construction workers who had just left or arrived for shift change, passersby, a group of medical personnel who were in training at the nearby Red Cross building, and a number of University of Minnesota students and staff. He said that 30–40 of these individuals went into the river to pull drivers and construction workers to safety.

Initial reports stated that the entire span of I-35W over the Mississippi River had collapsed while carrying bumper-to-bumper traffic and a full construction crew. About 6:10 p.m., the first Minneapolis Police Department unit arrived on scene. At 6:11 p.m., the first of 19 engine units from the Minneapolis Fire Department arrived. The first Hennepin County Sheriff's Office personnel arrived on the river at 6:14 p.m. to begin conducting the search and rescue of numerous people reportedly trapped in their vehicles in the river. The Hennepin County Medical Center also initiated its disaster plan, which involved calling in additional medical personnel, notifying other local hospitals, and dispatching all available ambulances to the scene.

About 6:10 p.m., a unified command post was established in the parking lot of the Red Cross building, located on the south side of the river just west of the area of the collapse.

The sheriff's office established its river incident command at 6:25 p.m. near the University of Minnesota River Flats area along the north bank of the river. This site had historically been used as a base for water rescue operations and was the nearest facility that had an adequate number of boat ramps for rescue operations. Within the first hour of the collapse, 12 other public safety agencies responded with 28 watercraft to assist with river rescue operations.

About 7:27 p.m., the fire department incident commander and the sheriff's office river operations incident commander decided to change the water operations from rescue mode to recovery mode.

Incident Command

The city of Minneapolis and Hennepin County use the unified command system in which the type of response required for an incident determines who will serve as the incident commander. In this accident, the assistant fire chief of the Minneapolis Fire Department was the incident commander, and the fire department was the lead agency responsible for overall operations as well as for issues related to the structural collapse of the bridge. The Minneapolis Police Department was responsible for the on-scene investigation (landside operations) and scene security, the Hennepin County Sheriff's Office was responsible for river rescue and recovery[2] (waterside operations), and the Hennepin County Medical Center ambulance service was in charge of emergency medical services operations.

On August 2, 2007, the U.S. Coast Guard established a temporary security zone on the Mississippi River from the Upper St. Anthony Falls Lock and Dam to the Lower St. Anthony Falls Lock and Dam. Access through this portion of the river was granted to emergency vessels only. About 7:00 p.m. on August 2, the Minneapolis Fire Department transferred incident command to the Minneapolis Police Department because the area had been declared a crime scene (because of the possibility that the bridge had been the target of a terrorist attack). The Hennepin County Sheriff's Office continued to be responsible for coordinating public safety dive teams searching the area around the bridge collapse and for using side-scanning sonar to attempt to locate vehicles and victims reported missing through Monday, August 6.

On Saturday, August 4, 2007, the sheriff's office requested and received the help of the Federal Bureau of Investigation (FBI) underwater search and evidence response team (USERT) and the U.S. Naval Sea Systems Command (NAVSEA) mobile diving and salvage teams. On August 5, USERT arrived and began river recovery operations. On August 6, Navy teams arrived and, along with USERT, assisted sheriff's office personnel, who continued to coordinate all water recovery operations until the last victim was recovered.

On September 6, 2007, river access was increased to allow limited commercial barge traffic. The Mississippi River was completely reopened for all river traffic on October 6.

---

[2] By Minnesota State statute (MSS 86B801), county sheriffs are responsible for recovering bodies from any waterway within their counties.

Injuries

A total of 190 people (vehicle occupants, construction workers, and Mn/DOT personnel) were confirmed to have been on or near the bridge when the collapse occurred. Medical records obtained as part of this accident investigation indicated that 145 people were transported to or treated at 12 area hospitals, medical centers, and clinics. The information in table 1 is based on data provided by the Hennepin County medical examiner and the treating hospitals.

Table 1. Injuries.

Injuries[A]	Total
Fatal	13
Serious	34 documented
Minor	111 (70 documented)
None or unknown	32
Total	190
[A]Title 49 *Code of Federal Regulations* (CFR) 830.2 defines a fatal injury as any injury that results in death within 30 days of the accident. A serious injury is defined as any injury that requires hospitalization for more than 48 hours, commencing within 7 days of the date the injury was received; results in a fracture of any bone (except simple fractures of the fingers, toes, or nose); causes severe hemorrhages or nerve, muscle, or tendon damage; involves any internal organ; or involves second- or third-degree burns, or any burns affecting more than 5 percent of the body surface.	

Accident Location

The accident occurred in the city of Minneapolis in Hennepin County, Minnesota. The I-35W bridge was located about 1 mile northeast of the junction of I-35W with Interstate 94. (See figure 2.) In addition to spanning the Mississippi River, the bridge also extended across Minnesota Commercial Railway railroad tracks and three roadways: West River Parkway, 2nd Street, and the access road to the lock and dam.

Bridge Description

General

The I-35W bridge (National Bridge Inventory structure no. 9340) was designed by the engineering consulting firm of Sverdrup & Parcel and Associates, Inc., of St. Louis, Missouri, a predecessor company of Sverdrup Corporation, which was acquired in 1999 by Jacobs Engineering Group, Inc. The bridge design was developed over several years, with plans for the foundation approved in 1964 and final design plans certified by the Sverdrup & Parcel project manager (a registered professional engineer) on March 4, 1965. The bridge design plans were approved by Mn/DOT on June 18, 1965.[3]

---

[3] For more information about revisions to the initial bridge design, see "Design History of I-35W Bridge" later in this report.

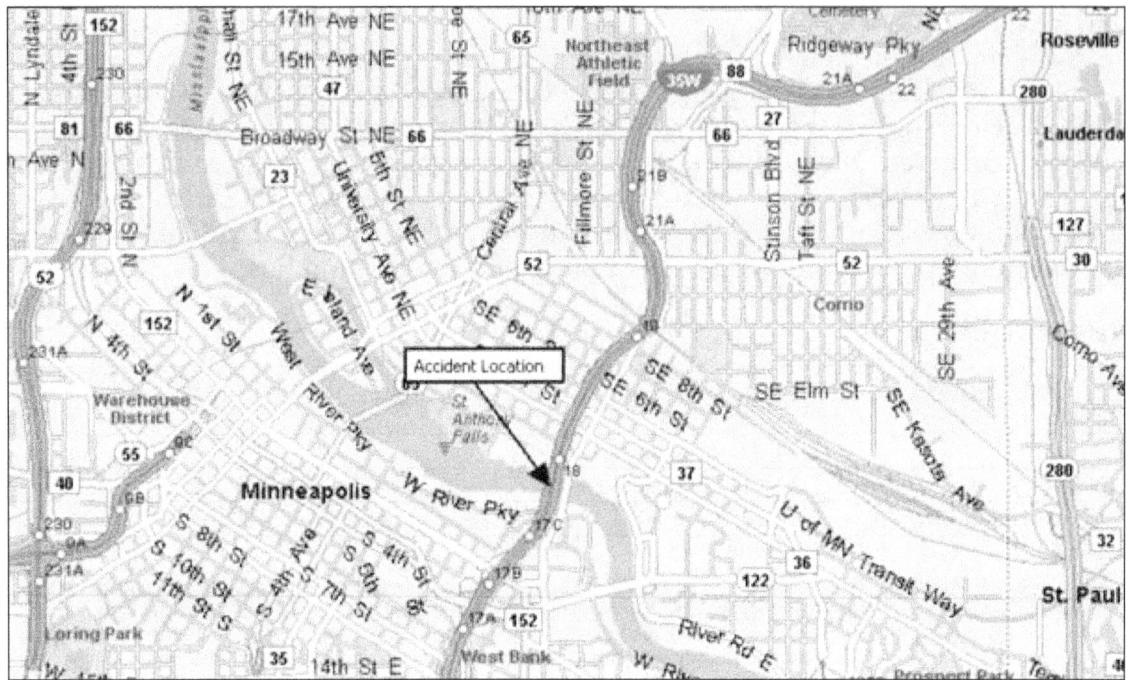

Figure 2. Accident location.

Construction of some piers started in 1964, and the bridge was opened to traffic in 1967.[4] The design was based on the 1961 American Association of State Highway Officials (AASHO)[5] *Standard Specifications for Highway Bridges* and 1961 and 1962 *Interim Specifications*, and on the *1964* Minnesota Highway Department *Standard Specifications for Highway Construction*. The bridge was built by Hurcon, Inc., with erection of the structure engineered and staged by the Industrial Construction Division of Allied Structural Steel Company, which was also the steel fabricator for the project. The structure used welded built-up steel beams for girders and truss members, with riveted and bolted connections. The bridge was 1,907 feet long and carried eight lanes of traffic, four northbound and four southbound. The structure inventory report for the bridge indicated average daily traffic in 2004 (the most recent available figures) as 141,000 vehicles. Average daily traffic of heavy commercial vehicles was 5,640. The earliest average daily traffic figures available for the bridge were from 1976 and indicated an average daily traffic at that time of 60,600 vehicles.

The bridge had 13 reinforced concrete piers and 14 spans, numbered south to north. (See figure 3.) Eleven of the 14 spans were approach spans to the deck truss portion. The bridge deck in the approach spans either was supported by continuous welded steel plate girders or by continuous voided slab concrete spans. The south approach spans were supported by the south abutment; by

---

⁴ The bridge was not opened to I-35W traffic until 1971, when construction of the I-35W highway was completed. Between 1967 and 1971, the I-35W bridge was used for traffic detoured from another bridge that was being renovated.

⁵ AASHO became AASHTO (American Association of State Highway and Transportation Officials) in 1973.

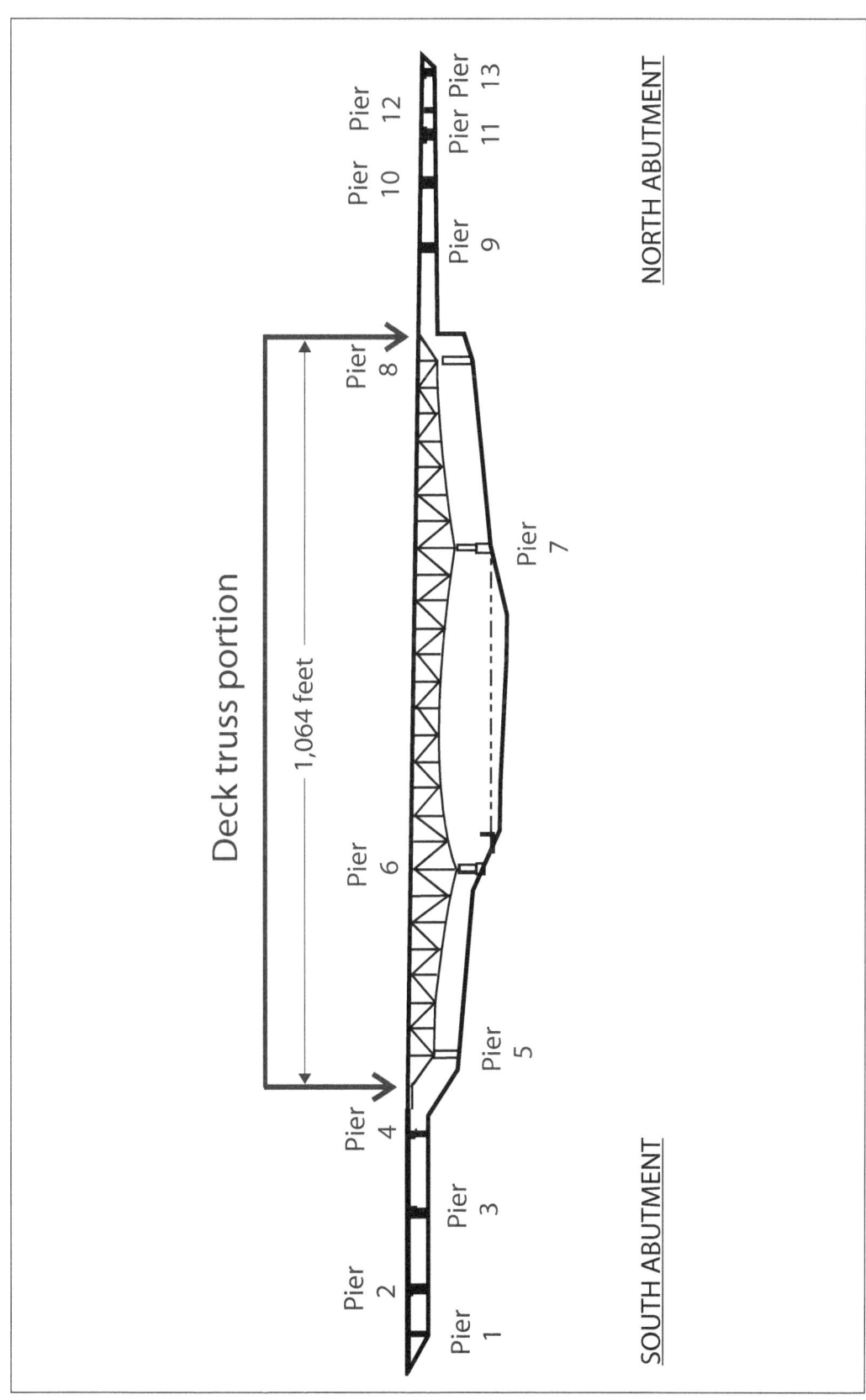

Figure 3. East elevation of I-35W bridge. The deck truss portion of the bridge extends from just south of pier 5 to just north of pier 8. (Source: Adapted from Mn/DOT graphic)

piers 1, 2, 3, and 4; and by the south end of the deck truss portion. The north approach spans were supported by the north abutment; by piers 9, 10, 11, 12, and 13; and by the north end of the deck truss portion. The 1,064-foot-long deck truss portion of the bridge encompassed a portion of span 5; all of spans 6, 7, and 8; and a portion of span 9. The deck truss[6] was supported by four piers (piers 5, 6, 7, and 8). (See figure 4.)

Figure 4. Center span of I-35W bridge, looking northeast. The center span is supported by pier 6 on the near (south) riverbank and pier 7 on the far (north) riverbank.

The original bridge design accounted for thermal expansion using a combination of fixed and expansion bearings for the bridge/pier interfaces. The fixed bearing assemblies were located at piers 1, 3, 7, 9, 12, and 13. Expansion (sliding) bearings were used at the south and north abutments and at piers 2, 4, 10, and 11. Expansion roller bearings were used at piers 5, 6, and 8. The roller bearings at pier 6 contained four large-diameter rollers; those at piers 5 and 8 were similar in design but contained only three large-diameter rollers. (See figure 5.)

---

[6] A *truss bridge* is typically composed of straight structural elements connected to form triangles. In a classical pin-jointed truss, the ends of the members are connected by pins that allow free rotation so the members only carry direct tension or compression with no bending. In large structures, pin joints are impractical, and members are joined with gusset plates. Geometry ensures that the members are still primarily loaded in direct tension or compression, but the gusset plate joints allow for transfer of some secondary bending stresses. Depending on the live load, the stress in some members may reverse, changing from tension to compression or vice versa. The I-35W bridge structure was in the form of a Warren truss with verticals. A Warren truss has alternating tension and compression diagonals and has the advantage that chord members are continuous across two panels, with the same force carried across both panels. This design simplifies, somewhat, the design calculations. In a *deck truss* bridge, the roadbed is along the top of the truss structure.

Figure 5. Expansion roller bearings at pier 5, looking east. (Source: URS Corporation)

The deck of the I-35W bridge consisted of two reinforced concrete deck slabs separated by about 6 inches. Each deck slab accommodated four 12-foot-wide traffic lanes and two 2-foot-wide shoulders. The deck slabs widened at the north end of the approach to accommodate on and off ramps and curved slightly at the south approach spans to match the roadway alignment. The bridge deck had 11 expansion joints.

The total width of the deck slabs was about 113 feet 4 inches. The northbound and southbound traffic lanes (four in each direction) were each about 52 feet wide. The northbound lanes were separated from the southbound lanes by a 4-foot-wide median. The railing along the outside edges of the bridge was about 2 feet 8 inches wide. When the bridge was opened to traffic in 1967, the cast-in-place concrete deck slab had a minimum thickness of 6.5 inches. As discussed later in this report, bridge renovation projects eventually increased the average thickness of the concrete deck by about 2 inches.

Deck Truss Portion

The deck truss portion of the bridge comprised two parallel main Warren-type trusses (east and west) with verticals. The upper and lower chords of the main trusses extended the length of the deck truss portion of the bridge and were connected by straight vertical and (except at each end of the deck truss) diagonal members that made up the truss structure. The upper and lower chords were welded box members, as were the diagonals and verticals, designed primarily for compression. The east and west sides of the box members are referred to as the side plates, and the top and bottom as the cover plates. The vertical and diagonal members designed primarily for tension were H members consisting of flanges welded to a web plate.

Riveted steel gusset plates at each of the 112 nodes (connection points) of the two main trusses tied the ends of the truss members to one another and to the rest of the structure.[7] The gusset plates were riveted to the side plates of the box members and to the flanges of the H members. All nodes had at least two gusset plates, one on either side of the connection point.[8] A typical I-35W main truss node, with gusset plates, is shown in figure 6.

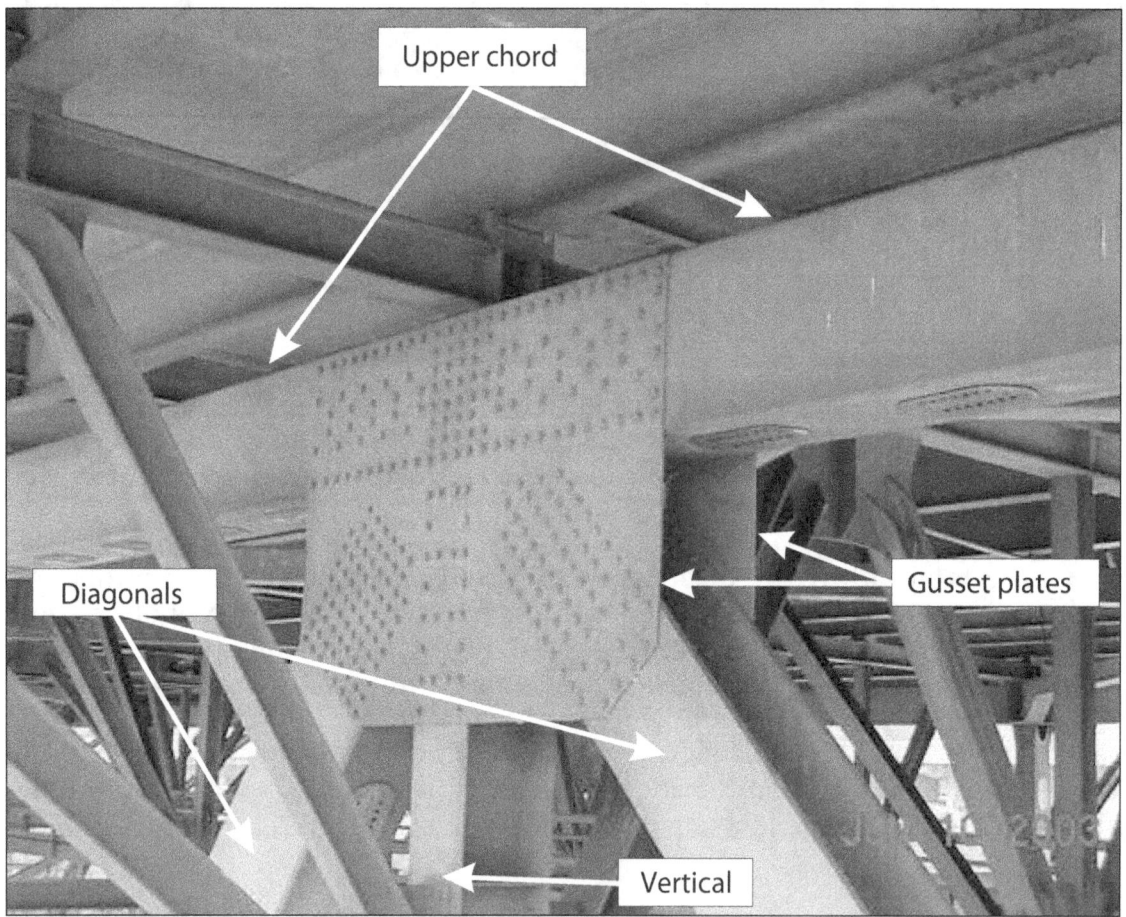

Figure 6. Typical five-member node (two upper chord members, one vertical member, and two diagonal members) on I-35W bridge. (Source: URS Corporation)

---

[7] At some gusset locations, bolts were used in place of some rivets, apparently to facilitate installation.

[8] At some locations, the configuration of the connection required two overlapping gusset plates on each side of the node.

The east and west main trusses were spaced 72 feet 4 inches apart and were connected by 27 transverse welded floor trusses spaced on 38-foot centers along the truss and by two floor beams at the north and south ends. The floor trusses were cantilevered out about 16 feet past the east and west main trusses. The concrete deck for the roadway rested on 27-inch-deep wide-flange longitudinal stringers attached to the transverse floor trusses and spaced on 8-foot-1-inch centers. (See figure 7.) The 14 stringers were continuous for the length of the deck truss except for five expansion joints. Diaphragms (generally C sections except at expansion joints, where they were wide-flange sections) connected the webs of adjacent stringers to transfer lateral loads and maintain structural rigidity and geometry. The ends of these diaphragms were riveted or bolted to vertical plates, called stiffeners, which were welded to and projected from the stringer webs.

Figure 7. Interior structure of north portion of deck truss, looking north. (Source: URS Corporation)

The bridge design documents and inspection records identified the 112 main truss nodes (29 upper and 27 lower nodes[9] on each of the two main trusses) by number. Because the bridge was longitudinally symmetrical (the north and south halves were almost mirror images), the nodes were numbered from each end starting at 0 on the south end and at 0' (0 "prime") on the north end. The node numbers increased from each end until reaching node 14, which was at the center of the bridge. Thus, *n* and *n'* identified the corresponding nodes on the bridge's south and north halves, respectively. The letters *U* (for upper chord) and *L* (for lower chord) further identified the nodes, along with *E* or *W* to specify the east or west main truss. For example, the "U5E" node refers to the sixth node from the south end of the bridge (because the node number begins at 0) on the upper chord of the east main truss. The "U5'E" node refers to the corresponding node on the north half of the bridge. (See figure 8.)

These node numbers were also used to identify the connecting main truss structural members. For example, the upper chord member that connected the U7 node to the U8 node was designated U7/U8. The vertical member that connected the U8 node to the L8 node was designated U8/L8. The diagonal members were similarly designated. Thus, the two diagonal members that extended from the U8 node to the two nodes on the lower chord were L7/U8 and U8/L9. Again, *E* and *W* were used to designate the east or west main truss.

The upper chord of each transverse floor truss was supported on and connected to the upper chords of the main trusses at the like-numbered nodes. For example, the floor truss that connected the main truss U10E and U10W nodes was designated floor truss 10.

The I-35W bridge was designed and built before metal fatigue cracking in bridges was a well-understood phenomenon. In the late 1970s, when a better understanding of metal fatigue cracking was established within the industry, deck truss bridges such as the I-35W bridge were recognized as being "non-load-path-redundant" — that is, if certain main truss members (termed "fracture-critical") failed, the bridge would collapse. According to Federal Highway Administration (FHWA) 2007 data, of the 600,000 bridges in the National Bridge Inventory, 19,273 are considered non-load-path-redundant. About 465 bridges within the inventory have a main span that is a steel deck truss.

Construction of Deck Truss

The method and sequence of construction of the deck truss portion of the bridge were specified in a series of engineering drawings. The span lengths required falsework (temporary support) below the trusses between piers 5 and 6 and between piers 7 and 8. Following erection of the main trusses between their south ends and pier 6 and their north ends and pier 7, the trusses were erected

---

[9] The 0 and 0' nodes were upper nodes only.

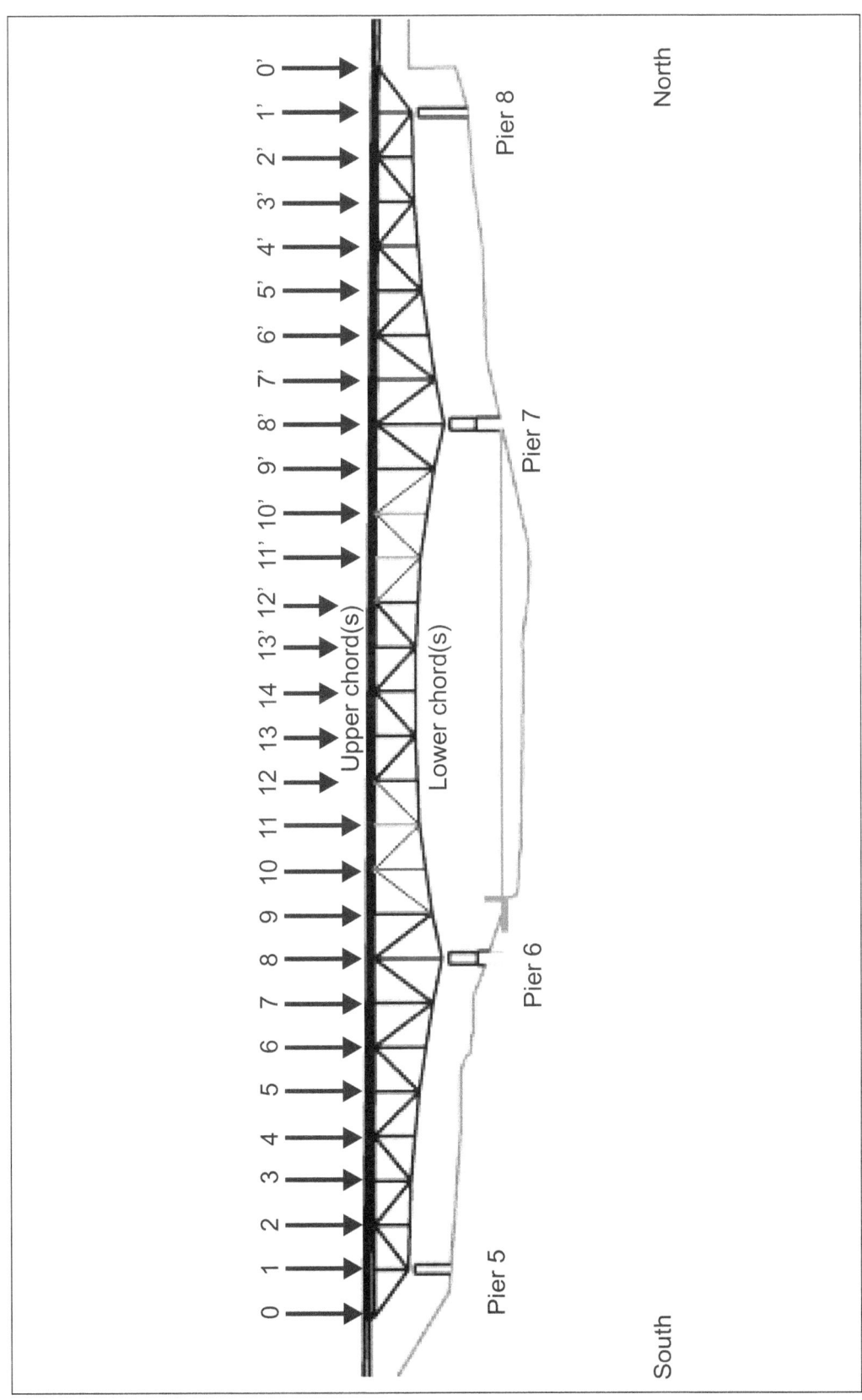

Figure 8. Nodes (connection points) on main trusses of I-35W bridge, designated by number. The nodes were further referenced by "U" for upper nodes (connecting members at the upper chord) and "L" for lower nodes (connecting members at the lower chord). Location of nodes on the east or west main truss was designated by "E" or "W," respectively. (Source: Adapted from Mn/DOT graphic)

toward the midpoint of the center span, with each half cantilevered from piers 6 and 7. The main trusses were then joined in the center, and final adjustments were made in the vertical positioning at piers 5 and 8.

Study of Collapse Video

As noted earlier, a motion-activated video surveillance camera at the Lower St. Anthony Falls Lock and Dam, just southwest of the bridge center span, activated as the collapse began. The camera captured a total of about 10 seconds (23 separate images) of the collapse sequence and saved it to a digital video recorder. FBI agents retrieved the recorder and forwarded it to the Safety Board's vehicle recorder laboratory in Washington, D.C., for analysis.

The primary structure visible in the video is the west truss of span 7, the bridge center span that crossed the river. Portions of the east truss are also visible behind the west truss. The visible section of the lower chord of the west truss extends from the L11W node, which is partially visible, northward through the L9'W node. The visible section of the upper chord extends from U11W northward through U8'W. Some upper nodes north of U8'W can also be seen, but they are generally too far away to be resolved clearly. Most of the west truss members are visible within this region. Most of pier 7 (on the north riverbank) can be seen, though a fence in the foreground obscures the west pier column. (See figure 9.)

Figure 9. View of a portion of the west side of I-35W bridge center span about 8 minutes before collapse. (Image extracted from surveillance camera video.)

The video does not show any element of the west truss south of vertical member U11/L11W or any element of the east truss south of the L13E node. None of the lower nodes on the west truss north of L9′W can be seen. Most of the east truss upper nodes that are visible to the camera are either too dark or blurred to be identified or are obscured by the west truss.

The video system activated at 18:04:57 central daylight time,[10] just after initiation of the collapse. The first video image of the collapse (see figure 10) shows that the southern end of the visible structure of span 7 has dropped downward as compared to its position in precollapse recorded images. In the north portion of the span, a bend appears in lower chord members L9′/L10′W and L10′/L11′W.[11] In the south portion, deck stringer 14, the westernmost stringer, has begun to separate from the deck at the U11W node. Over the next 3 seconds, the entire center span separates from the rest of the bridge structure and falls into the river. The video shows that during the collapse of this span, the bridge deck remained generally level east to west, but the north end of the span remained higher than the south end as the bridge section fell.

Figure 10. Close-up of first image of collapse sequence, captured by surveillance camera at 18:04:57. The south (near) end of the center span can be seen to have dropped from its position as shown in precollapse images (compare with figure 9).

---

[10] All times in this section reflect the U.S. Naval Observatory clock, offset to central daylight time.

[11] This chord section was actually one continuous member between the L9 W and L11′W nodes that passed through the L10 W node.

Pier 7, on the northern bank of the river, appears to have remained vertical and stationary until after the collapse began. Following the collapse, survey measurements indicated that pier 7 was tilted approximately 9° to the south.

Damages

Approach Spans

Although the primary damage occurred in the deck truss portion of the I-35W bridge, the approach spans also sustained damage in areas where the cantilevered ends of the spans had been supported by the ends of the deck truss. The damage sustained was consistent with a loss of this support, which caused the ends of the approach spans to drop. Although the steel girders in the approach spans contained previously documented fatigue cracks, no evidence was found that these cracks affected the damage patterns to these spans.

Deck Truss Spans

During the collapse sequence, the deck truss portion of the bridge separated into three large sections. Most of the center span, between piers 6 and 7 (referred to here as the "center section"), separated from the remainder of the truss and fell into the river. The section south of pier 6 (the "south section") fell onto land on the south side of the river, and the section north of pier 7 (the "north section") fell onto land on the north side of the river. (See figure 11.) General damage to each of these sections is detailed below. The fractures in the members between these sections are discussed later in this report in the "Examination of Deck Truss Fracture Areas."

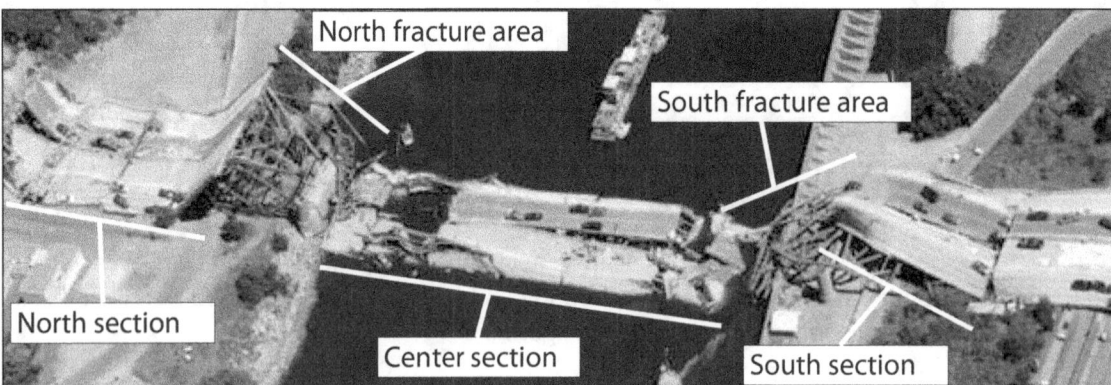

Figure 11. Collapsed deck truss sections of I-35W bridge.

South Section. Figure 12 shows the collapsed south section of the deck truss. The deck and floor stringers north of the U8 node (directly over pier 6) separated at this node and remained attached to the center section of the structure, which fell into the river. The deck and stringers south of the U8 node remained approximately in their original positions relative to the upper chords of the truss. The U8 and L8 nodes, east and west (at pier 6), remained with the collapsed south section of the truss.

Figure 12. Looking south at south fracture area and a portion of collapsed south section.

The L1 nodes were found north of pier 5, and the L8 nodes were found north of pier 6, indicating that, in general, the entire south section of the truss had displaced some amount north toward the river. Associated with this movement, the lower surface of lower chord members L7/L8 east and west contained impact and scraping damage from contact with the expansion rollers and upper surface of pier 6. Loading associated with this damage was severe enough that lower chord member L7/L8E was partially fractured at approximately mid-length. Additionally, member L7/L8E had broken away from the L7E node (through shearing of the rivets between the member side plates and the gusset plates) and the L8E node (through fracture of the member and shearing of the remaining rivets between the member side plates and the gusset plates).

The upper chords of the south section were intact between nodes 1 and 8. (The upper chord on both trusses was fractured between nodes 0 and 1.) The lower chord of the west truss was fractured between the L3W and L4W nodes, and the lower chord of the east truss was fractured adjacent to the L1 node. Also, as previously noted, each end of lower chord member L7/L8E separated from the L7 and L8E nodes, primarily through shearing of the gusset plate rivets.

In the postcollapse position, the portion of the south section between nodes 4 and 8 was toppled (laid over) to the east. Lower chord member L7/L8W remained on top of pier 6 west, while the L7E and L8E nodes struck the ground. Most of the main truss members and nodes in the south section contained compression or bending damage consistent with ground impact. Mating fracture areas in the lower chords were displaced, indicating continued translation to the north after the fractures were created in the lower chords.

The floor trusses from node 8 southward remained at least partially attached to the nodes on the east and west main trusses. The sway braces and lateral braces from nodes 0–8 showed no evidence of primary failure.

Center Section. See figure 13 for a photograph of the center span before collapse. In the postcollapse position, the center section of the truss was relatively flat and almost directly below its original location. (See figure 14.) As noted previously, the video of a portion of the collapse sequence captured by a nearby surveillance camera showed that the south end of the center section (just north of pier 6) dropped before the north end (just south of pier 7) and that the center section remained relatively level east to west as it fell. Many of the vehicles in the inner northbound lanes of the bridge remained in their lanes as the collapse occurred, indicating that the east and west main trusses at the south end of the center section fractured at about the same time and that the center section dropped into the river with a minimum of lateral roll.

The main truss upper chord members from nodes 12–12' remained intact and above the water, with the floor trusses at these nodes at least partially attached and supporting the stringers and deck. The video recording showed no failure or deformation occurring in the lower chords between the L12E and L12'E nodes and between the L11W and L12'W nodes. The L11W node appeared to remain intact during the collapse. Three of the five truss members that meet at the L11W node are visible in the video; and these members appeared to maintain their orientation relative to one another, within the resolution of measurement, as span 7 collapsed.

The lower chord members, lower lateral braces, and sway braces from the center section were generally under water. The fractured L9 node ends of the lower chord members L9/L10 east and west were resting against the lock guide wall near pier 6.

Figure 13. Center span of I-35W bridge (before collapse), looking northeast, with certain nodes labeled.

Figure 14. Collapsed bridge center section, looking southeast.

North Section. An overall view of the north portion of the deck truss postcollapse is shown in figure 15. A large portion of the deck truss above and north of pier 7 had rotated to the north as a rigid unit. This portion included the deck and upper chords from the U8' to U6' nodes, the lower chords from the L9' to L7' nodes, and the diagonals and verticals between these nodes. In the postcollapse position, lower chord members L7'/L8' east and west were contacting the north side of the upper end of the pier 7 columns. The video shows that the rigid portion of the deck truss initially remained in position and did not rotate as the center span dropped into the river. Later in the recording, after the water splash clears and the debris settles, this portion of the structure can be seen to have rotated to the north.

Figure 15. Collapsed north section of bridge.

Between this rigid body portion and pier 8, the deck truss had collapsed almost straight downward, resulting in severe vertical compression damage. In this area, the upper chords and upper nodes generally were displaced 18–25 feet to the north relative to the lower chord and lower nodes. The upper truss chords at nodes 2' were separated as a result of fractures of the gusset plates at these nodes.

The trusses, stringers, and deck at the U1' node (above pier 8) were bent down over the pier. The deck had fractured at this bend location, but the stringers were still mostly intact and had severe bending and lateral buckling. The L1' nodes were on the ground, resting against the north face of pier 8. The upper chord of floor truss 1' was on top of or on the north side of the pier, and the lower chord of the floor truss was on the south side of the pier. Nodes 0' east and west were on the north side of the pier. Based on the postcollapse position of floor truss 1' and the deck truss above pier 8, the truss collapsed across the pier without significant movement relative to the top of the pier. As the portion of the truss south of pier 8 dropped, the pier would have been pulled to the south, consistent with its cracking and tilt to the south.

Piers and Bearings

Piers 5 and 6 were minimally damaged during the collapse, and postcollapse survey measurements indicated that these piers exhibited no settlement or displacement. Piers 7 and 8 were both tilted about 9° south, toward the river, in their postcollapse positions. The video recording showed no evidence that pier 7 shifted before the initiation of the collapse, indicating that its movement occurred during the collapse. Excavation at these damaged piers showed that their tilted postcollapse positions occurred because of separations above the bases of the piers. Pier 7 (the fixed bearing location) hinged about the top of the pier footing, and the pier 8 columns hinged about a section approximately 3.5 feet above the top of the footings, at the termination of the footing dowel reinforcement. There was no evidence that the footings of these piers shifted.

The bearing rollers at piers 5 and 6 had come off the north side of the piers (with the exceptions of one roller that remained on pier 5 west and one roller that was found on the south side of pier 5 east). The wear patterns on the bearing sole plates indicated that the rollers had been moving annually by as much as 5 inches on pier 5 and 2.5 inches on pier 6, which was consistent with the design.

The wear patterns on the bearings at pier 8 showed evidence of normal movement over a distance of 2.5 inches, similar to the amount of movement for the rollers in pier 6. The bearing rollers at pier 8 had come off the south side of the pier (with the exception of one roller that was found on the north side of pier 8 west). The roller wear marks on the bearing plates at piers 5, 6, and 8 were approximately in the center of the plates, indicating that there was no significant longitudinal movement of the piers before the collapse.

Recovery of Structure

Initial removal of the collapsed truss structure from the accident site took place under the direction of the Hennepin County Sheriff's Office as part of the search for accident victims. Because finding and identifying victims had a higher priority than preserving evidence, some postaccident damage was done to the bridge structural components as they were removed from the scene. For example, during the initial phase of removal and before documentation of the preaccident locations of the deck stringers, the reinforced concrete deck and the stringers were cut or pulled apart with a shear-equipped backhoe, which caused additional damage to the stringers. The longitudinal locations of some of these stringers could be determined by the locations of shear studs along the top flange of the stringer ends. Also during this phase of the recovery, some larger members of the main truss were torch cut to permit barge access for victim recovery; these cuts were made under the guidance of Safety Board investigators.

Following this initial removal phase and in situ inspections and documentation, the removal of bridge components was directed by Mn/DOT, with the concurrence of Safety Board investigators. Individual pieces were assigned a salvage number and noted with the identification of the structural member, postaccident location, and date of removal.

Because much of the bridge center span collapsed into the river, its structural components were inaccessible for detailed inspection until after their removal from the site. During the earliest stages of recovery, some of these members were cut with a shear and pulled and twisted in an attempt to remove them. Later, less aggressive methods were employed, but some structural members were lifted out of the water while still connected to other members, and additional deformation may have occurred.

The recovered truss portions of the bridge were moved to Bohemian Flats Park, just downriver and east of the accident site, where the components were laid out in their relative original positions in the structure and subjected to detailed Safety Board examination. The results of those examinations are discussed later in this report.

Bridge Renovations and Modifications

Between the time of its opening in 1967 and the accident, the I-35W bridge underwent three major renovation/modification projects, two of which increased the dead load[12] on the structure.[13] The following sections provide details of those projects.

---

[12] *Dead load* refers to the static load imposed by the weight of materials that make up the bridge structure itself.

[13] As part of another project in 1999, portions of the deck truss were painted, and coverings (bird screens) were installed over the openings in the box members.

1977 Renovation: Increased Deck Thickness

The 1977 construction plan for renovation of the I-35W bridge involved milling the bridge deck surface to a depth of 1/4 inch and adding a wearing course of 2 inches of low-slump concrete.[14]

According to Mn/DOT representatives, the bridge was initially constructed with 1.5 inches of concrete cover over the uncoated top reinforcing bars (rebar) in the bridge deck. By the early 1970s, States with a similar thickness of covering over rebar and with severe operating environments, like Minnesota, were experiencing rebar corrosion. To combat this problem, Mn/DOT adopted a policy of using an additional layer of concrete overlay to increase to 3 inches the cover over the upper layer of deck rebar. According to Mn/DOT officials, this additional cover reduced the likelihood that harsh road chemicals could reach and react with the steel rebar, thus substantially extending the life of bridge decks. Based on measurements after the accident, this project increased the average deck thickness to approximately 8.7 inches. The additional concrete applied as part of this project increased the dead load of the bridge by more than 3 million pounds, or 13.4 percent.

1998 Renovation: Median Barrier, Traffic Railings, and Anti-Icing System

Work done on the I-35W bridge in 1998 involved replacing the median barrier, upgrading outside concrete traffic railings, improving drainage, repairing the concrete slab and piers, retrofitting cross girders, replacing bolts, and installing an anti-icing system.

Mn/DOT representatives stated that several features of the bridge, including the median barrier and outside traffic railings, did not meet then-current (1998) safety standards. Also, the original median barrier and traffic railings were deteriorating from corrosion and traffic impact. The permanent changes to the median barrier and railings increased the dead load on the bridge by about 1.13 million pounds, or 6.1 percent. During the 1998 construction project, temporary barriers were erected in both the northbound and southbound lanes to protect workers in the median and to protect the new median barrier while it was curing. The temporary barriers were then moved to outside lanes to protect workers while the outside traffic railings were upgraded. The two rows of temporary barriers weighed 1.60 million pounds (840 pounds per foot), which was evenly distributed along the 1,907-foot length of the bridge.

The I-35W bridge was a candidate for installation of an anti-icing system because of the high incidence of winter traffic accidents on the bridge. The anti-icing system worked with a combination of sensors, a computerized control system, and a series of 38 valve units and 76 spray nozzles that applied potassium acetate to the roadway.

---

[14] *Slump* refers to the consistency of the concrete and is measured in inches. Concrete with a lower ratio of water in the mix will have less slump than concrete with a high water content. According to Mn/DOT, the acceptable slump for a (low-slump) concrete overlay is 3/4 inch versus about 4 inches for other parts of a bridge deck.

2007 Repair and Renovation: Repaving

Roadway work was underway on the I-35W bridge at the time of the collapse. This work—which was being performed by Progressive Contractors, Inc. (PCI), of St. Michael, Minnesota—involved removing the concrete wearing course to a depth of 2 inches and adding a new 2-inch-thick concrete overlay. The construction plan also called for removing unsound concrete from the curb and patching it with concrete, reconstructing the expansion joints, and removing and replacing the anti-icing system spray disks and sensors in the deck. The project began in June 2007 and had a scheduled substantial completion date of September 21, 2007, with final completion expected by October 26, 2007. PCI had completed seven pavement section overlays and was preparing for the eighth section overlay pour when the bridge collapsed.

At the time of the bridge collapse, four of the eight travel lanes (the two outside lanes northbound and the two inside lanes southbound) were closed to traffic. The preexisting wearing surface was still in place on the two inside lanes northbound, where the average deck thickness was 8.7 inches. The new overlay was already in place on the two outside northbound lanes (average deck thickness of 8.8 inches) and the two outside southbound lanes (average deck thickness of 8.9 inches). The surface of the two inside southbound lanes had been milled for the entire length of the bridge, removing about 2 inches of material. The weight of the concrete that had been milled from the two inside southbound lanes in preparation for the overlay was calculated by the FHWA to be 585,000 pounds, of which about 250,000 pounds would have come from the bridge center span.

Staging of Construction Materials on Bridge Deck

Of the seven overlays that PCI had completed on the I-35W bridge before the collapse, five involved staging construction materials, primarily aggregates, on the bridge ramps; one involved placing materials on the bridge deck; and one involved staging materials on both the ramps and the deck.

The staging of aggregates and other construction materials on a bridge is sometimes considered necessary to allow timely delivery of concrete to the job site. For example, because the low-slump concrete used for the overlay has a lower water and higher cement content than typical concrete, it sets up very quickly after mixing. Minnesota specifications require that all concrete overlays be mixed at the job site, with a 1-hour window from initial mixing to final concrete screeding, or setting of the final grade, and a 15-minute window from initial deposit of concrete on the deck to final screeding. Because of the requirements for quick placement of concrete and the relatively low volumes of concrete needed for overlay pours, the use of ready-mix concrete trucks is not practical; and, according to Mn/DOT, ready-mix concrete has not been used on State bridges. In the case of the I-35W bridge, storing materials off the bridge deck and thus farther from the job site would have required that small motorized buggies be used to transport the mixed concrete several hundred feet and might have required closing additional traffic lanes.[15]

The first time PCI staged construction materials on the I-35W bridge deck (though not on the deck truss portion of the bridge) was on the night of July 6, 2007. A portion of the materials for a projected 750-foot overlay on the outside southbound lanes from pier 6 northward toward pier 8 was staged on the bridge deck, centered approximately over the midpoint between north approach spans 10 and 11. This staging of equipment and aggregates was similar to the arrangement on the day of the accident except for two additional loads of sand and two additional loads of gravel that were off the bridge.

According to PCI, vehicles and aggregates were staged on the deck truss of the bridge for a pour that began on July 23, 2007. The overlay was a planned 589-foot pour on the northbound lanes extending from near the midpoint of span 8 just south of pier 8 (approximately at node 4') to the expansion joint at the end of span 7 (approximately at node 8). The staging area for this pour was about 183 feet long and extended from node 4' to the north end of the deck truss. The staging area had three 24-ton loads of gravel and three 24-ton loads of sand. PCI representatives told the Safety Board that a Mn/DOT bridge construction inspector was on scene for each pour.

According to Mn/DOT, bridge construction inspectors are assigned to a project to ensure that the contractor fulfills its responsibilities and that all materials used meet required standards. On the I-35W bridge project, this would include checking to see that the concrete met Mn/DOT specifications. Bridge construction inspectors are not engineers, nor are they trained in bridge inspections. In a postaccident interview, the PCI job foreman (who had worked for the company for 21 years and had been on the I-35W bridge project for 3 weeks) told the Safety Board that he had asked a Mn/DOT bridge construction inspector if materials could be staged on the bridge for the July 23 pour in the northbound lanes. He said the inspector evidenced no concern about the staging, which the job foreman interpreted as permission. The foreman said the reason he asked was because of the time and labor that would be required to move materials and clean the area after delivery. He did not indicate that he considered the weight of the materials to be an issue.

On the afternoon of August 1, 2007, PCI was preparing to pour a 530-foot overlay in the southbound inside lanes. The pour, which was not to begin until 7:00 p.m. because of the day's high temperature, would extend between node 14 near the center of the deck truss northward to node 0'. The PCI job foreman calculated and ordered the materials that would be needed for the job. The order included four end-dumps of sand, four end-dumps of gravel, and three cement tankers. Two of the cement tankers were fully loaded 80,000-pound vehicles

---

[15] Mn/DOT representatives told the Safety Board that during the repaving of three other State bridges similar in length to the I-35W bridge, the concrete overlay material was mixed off the bridge in the approach roadway right-of-way and transported to the job site in buggies.

that would be positioned off the deck. One tanker was of legal weight (less than 80,000 pounds) and would be staged on the deck.

The aggregates and associated construction vehicles would come to occupy an estimated 228-foot-long section of the two closed southbound lanes over the main river span (span 7). The area where the equipment and materials were stored was established by interviewing witnesses, by examining the limits of the work area, and by examining photographs, including one (figure 1 top) taken about 3:45 p.m. by a passenger aboard a commercial airliner. After this photograph was taken (see figure 16), PCI workers moved the materials closer to the median barrier to allow more room for the movement of construction vehicles and traffic.

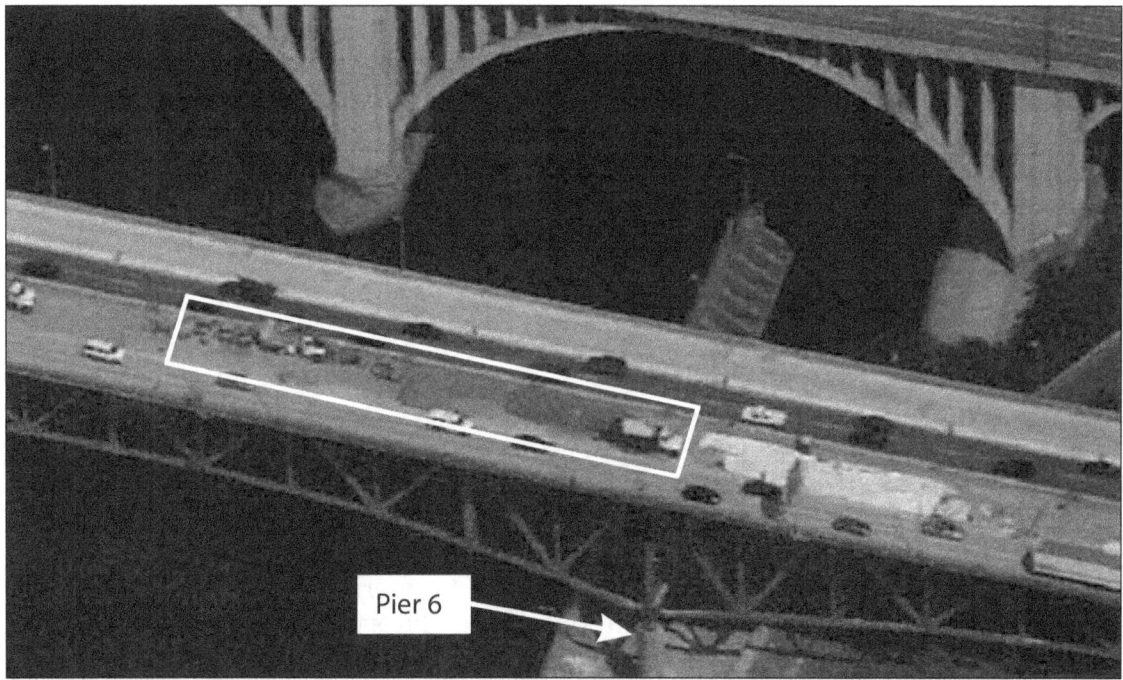

Pier 6

Figure 16. Construction equipment and aggregates (indicated by white box) stockpiled on southbound lanes of bridge about 2 hours 15 minutes before collapse. (This figure is an enlargement of a section of a photograph taken by a passenger in a commercial airliner departing Minneapolis/ St. Paul International Airport.)

The construction aggregates were distributed in eight adjacent piles (four sand and four gravel) placed along the median in the leftmost southbound lane just north of pier 6. The combined aggregates occupied a space about 115 feet long and 12–16 feet wide, with its southern boundary about 10 feet north of pier 6. (See figure 17.) This staging placed the aggregate piles generally centered longitudinally over the deck truss U10 nodes. Along with the aggregates in this area were a water tanker truck with 3,000 gallons of water, a cement tanker, a concrete mixer, one small loader/excavator, and four self-propelled walk-behind or ride-along buggies for moving smaller amounts of materials.

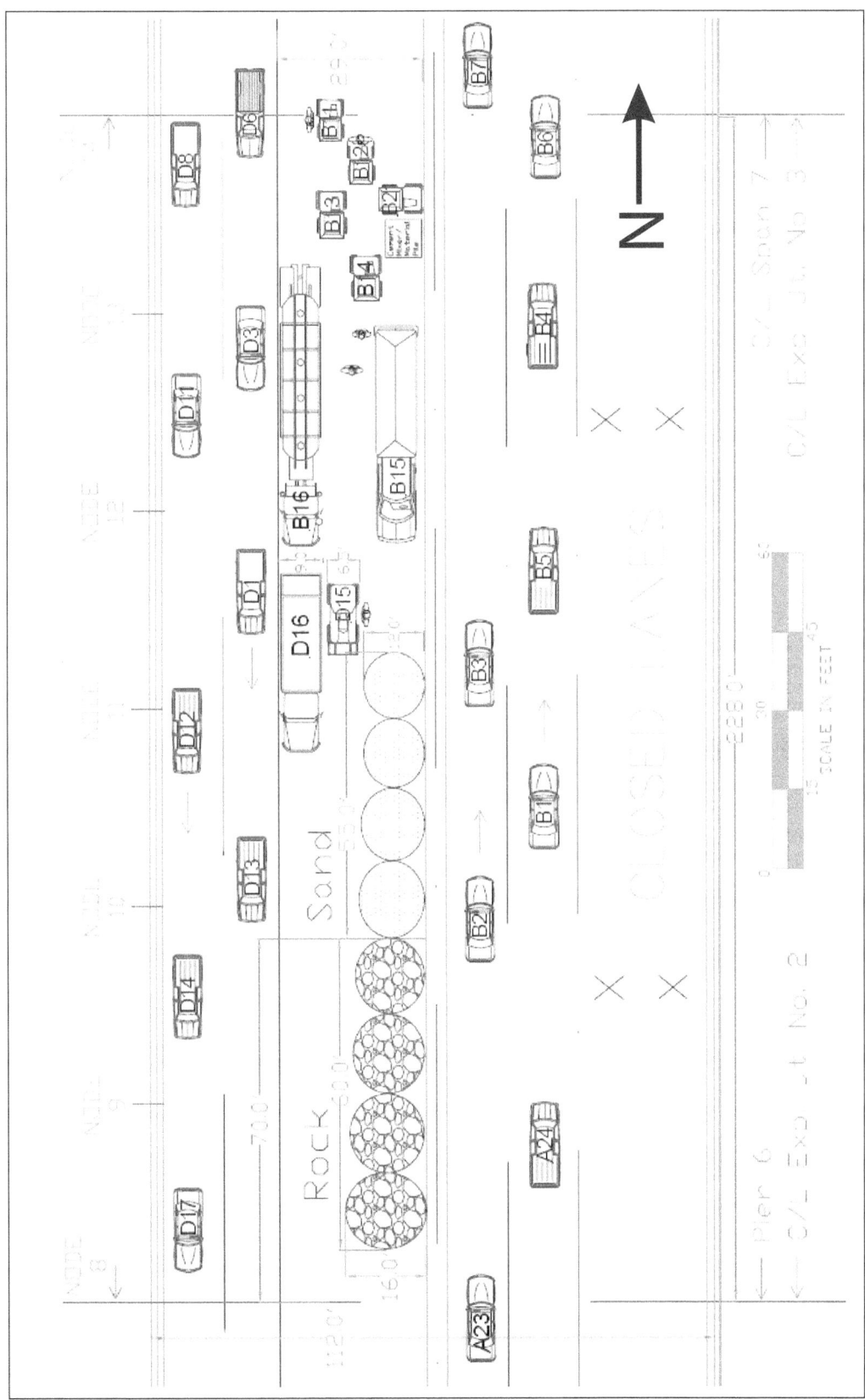

Figure 17. Reconstruction of estimated positions of construction aggregates and equipment along the two inner southbound lanes of I-35W bridge at time of collapse.

Bridge Traffic and Construction Loading at Time of Collapse

The load on the I-35W bridge at the time of the collapse included vehicular traffic as well as temporary loads associated with the construction work on the bridge. The documented delivered weights for the aggregates staged on the bridge were 184,380 pounds of gravel and 198,820 pounds of sand, for a total weight of 383,200 pounds. The estimated weight of the parked construction vehicles, equipment, and personnel in this area was 195,535 pounds, for a total estimated weight of 578,735 pounds positioned over the inner west side of the bridge center span just north of pier 6.

Based on postaccident vehicle positions, photographs, and witness statements, investigators were able to determine the types and general positions of the 111 vehicles (including construction vehicles) on the bridge at the time of the collapse. The positions of the 17 construction workers on the bridge were also documented.

The vehicles that were on the bridge were weighed either during removal from the accident site or later at an impound lot. These weights were adjusted for occupants and cargo and, in some cases, for water and debris. Weights were also estimated for the various pieces of construction and related equipment and for construction personnel. These weights were totaled to arrive at the approximate traffic and construction loads shown in table 2 for various points along the bridge at the time of the collapse.

Table 2. Summary of approximate loads along I-35W bridge at time of collapse.

Lanes	South of pier 6 (pounds)	Center span between piers 6 and 7 (pounds)	North of pier 7 (pounds)	Total (pounds)
Open southbound	112,200	64,650	98,050	274,900
Closed southbound	41,900	578,735	91,691	712,326
Open northbound	66,300	57,100	44,950	168,350
Closed northbound	104,750	0	0	104,750
Total	325,150	700,485	234,691	1,260,326

Mn/DOT Policies Regarding Construction Loads on Bridges

Mn/DOT representatives told the Safety Board that any questions regarding the stockpiling of materials on the bridge should have been formally directed to the project engineer (in writing) rather than orally directed to a construction inspector. Mn/DOT's *Standard Specifications for Construction* 1509 and 1510, which are part of the contract documents for every project, address the respective authorities and duties of the project engineer (1509) and construction inspectors (1510). Specification 1509 stated:

> The Project Engineer is the engineer with: (1) Immediate charge of the engineering details of the construction Project. (2) Responsibility for the

administration and satisfactory completion of the Project. (3) Authority commensurate with the duties delegated to the Engineer. (4) Authority to reject defective material and to suspend any work that is being improperly performed.

Specification 1510 stated:

Inspectors employed by the Department [Mn/DOT] will be authorized to inspect all work done and materials furnished. The inspectors will not be authorized to alter or waive the provisions of the Contract, to issue instructions contrary to the Contract, or to act for the Contractor.

As a representative of the Engineer, the inspector will report progress and acceptability of the work being performed, and will call to the attention of the Contractor any failures and infringements on the part of the Contractor. Should any dispute arise as to the materials or work performance, the inspector may reject materials and suspend operations until the question at issue can be referred to and be decided by the Engineer.

The Mn/DOT project engineer for the ongoing work on the I-35W bridge stated that he typically made three or four trips per week to the bridge and relied heavily on the experience of the Mn/DOT project construction supervisor to bring potential problems to his attention.

The project construction supervisor had worked for Mn/DOT for more than 20 years and had been the inspector for one previous large bridge overlay job. He said that on one occasion during the I-35W bridge project, the contractor inquired about using the pavement planing/milling machine to go below the standard 2-inch depth to remove larger sections of unsound concrete. The construction supervisor referred the question to the Mn/DOT Bridge Office, which—after conducting an experiment at another location—determined that the milling should not exceed 2 inches because of concerns that the vibration might damage the structure.

The construction supervisor told Safety Board investigators that he had had many contractors approach him about placing heavy machinery on bridges over the years but that he had never been asked about placing heavy piles of aggregates. He said, however, that most of the more than 60 projects he had worked on involved shorter bridges where there would have been no need to store materials on the bridge itself. He said that he was not at work on August 1 and was not aware of the stockpiling of materials. Asked whether he would have objected to the stockpiling had he known about it, he said:

I'm not sure. I would have had to look at the loads. My best guess is it would have been a 50-50 chance that I might have done something, and that is only because of my close working relationship with the engineers in the bridge office and my many years of experience.

A Mn/DOT bridge construction inspector was on scene when the materials were delivered and placed on the bridge and was standing near the stockpiled materials when the bridge collapsed.

According to Mn/DOT, the agency's policies at the time of the accident did not specifically address the placement of construction aggregates on bridges. Mn/DOT told the Safety Board:

> The contractor can request to place larger than legal loads on a new or remodeled bridge with Mn/DOT Construction Project Engineer's approval. Although not a written policy, when a contractor proposes a load that exceeds legal loads, it is a practice for the Mn/DOT Construction Project Engineer to consult with the Regional Construction Engineer in the Bridge Office. The construction loading information is provided to the Load Rating Unit or Design Unit for evaluation to determine if the loading is acceptable or if any special procedures such as use of the load distribution mats are required. Some examples of loads that exceed legal loads are mobile cranes or heavy earth moving equipment.

Mn/DOT policy regarding construction loads on bridges is contained in its *Standard Specifications for Construction*. At the time of the collapse, section 1513 of the specifications, "Restrictions on Movement of Heavy Loads and Equipment," read, in part:

> The Contractor shall comply with legal load restrictions, and with any special restrictions imposed by the Contract, in hauling materials and moving equipment over structures, completed upgrades, base courses, and pavements within the Project that are under construction, or have been completed but have not been accepted and opened for use by traffic.
>
> The Contractor shall have a completed Weight Information Card in each vehicle used for hauling bituminous mixture, aggregate, batch concrete, and grading material (including borrow and excess) prior to starting work. This card shall identify the truck or tractor and trailer by Minnesota or prorated license number and shall contain the tare, maximum allowable legal gross mass, supporting information, and the signature of the owner. Equipment mounted on crawler tracks or steel-tired wheels shall not be operated on or across concrete or bituminous surfaces without specific authorization from the Engineer. Special restrictions may be imposed by the Contract with respect to speed, load distribution, surface protection, and other precautions considered necessary.
>
> Should construction operations necessitate the crossing of an existing pavement or completed portions of the pavement structure with equipment or loads that would otherwise be prohibited, approved methods of load distribution or bridging shall be provided by the Contractor at no expense to the Department.

Neither by issuance of a special permit, nor by adherence to any other restrictions imposed, shall the Contractor be relieved of liability for damages resulting from the operation and movement of construction equipment.

A postaccident report from Mn/DOT to the Safety Board stated:

Had this proposal [to stage materials on the bridge] been forwarded to us from the contractor at the start of the overlay contract, we would likely have rejected it, before doing any analysis for the loads. We would have questioned if there were alternate locations for stockpiling the materials. This loading is immediately seen to be much larger than design loads. For example, the HS 20 design lane load is 0.64 k/ft [640 pounds per foot]. The rock and sand piles weigh about four times as much as this, spread over a width of 14 ft., just slightly more than a design lane. [This comparison to design load pertains to loading directly below the stockpiled materials, not to loading on the entire span.]

PCI representatives told the Safety Board that

To meet the tight deadlines and being forced to operate within specific traffic configurations posed unique and very difficult obstacles. With this in mind, PCI on numerous occasions requested additional lane closures and additional full weekend closures. On one occasion we even formally requested to batch the wearing course material out of our nearby PCI plant site. These requests were denied.

At the request of the Safety Board, Mn/DOT representatives performed a load rating analysis to determine whether, based on documented design calculations, the construction loads did, in fact, exceed allowable load levels for the bridge structural members. This load rating was done in accordance with the AASHTO *Bridge Design Specifications*, 17th edition, and the *Manual for Condition Evaluation of Bridges,* 2nd edition.

The construction and materials loads were modeled and analyzed using a combination of tools including Bridge Analysis and Rating System (BARS) software. The bridge dead loads used in the analysis were increased by 19.5 percent over the original 1967 loads to account for the load increases from the 1977 and 1998 bridge repair and renovation projects. The dead load values were not reduced to account for the 2 inches of concrete that had been milled from two travel lanes before the collapse.

The BARS software was used for analysis of stringers. The floor trusses and main truss members were analyzed using the original Sverdrup & Parcel calculations, original truss influence lines, and hand calculations.

The analysis assessed the effect of the additional loads on the most affected structural members, which were determined to be two deck stringers, two floor trusses, and nine main truss members. The analysis concluded that, for all the structural members considered, the construction loads were within operating limit capacities — which means that the structural members complied with the AASHTO bridge design specification requirements for strength under the construction loads plus full HS20 vehicle traffic design loading on the four lanes open to traffic during this construction phase.

The analysis considered only the primary truss members and did not consider the gusset plates joining the truss members. BARS was not used for the truss members because it can only analyze a simple span truss, and the program had no provision for including gusset plates. In late 2007, Mn/DOT was in the process of phasing out the BARS program and replacing it with Virtis, a specialized bridge load rating and design analysis program developed by the FHWA and AASHTO. Like the BARS program, Virtis does not have the capability to include gusset plates in the analysis.[16] Safety Board investigators were told by those with experience in the bridge industry that the design methodology for gusset plates is normally considered "very conservative," with the result that a gusset plate is generally assumed to be stronger than the beams it connects. Additionally, all the State transportation officials contacted by the Safety Board in conjunction with this accident investigation indicated that they considered gusset plates to be stronger than the members they connect.

On August 8, 2007, the FHWA issued Technical Advisory 5140.28, *Construction Loads on Bridges*, which states

> While no conclusions have been reached, in an abundance of caution, we strongly advise the State Transportation Agencies and other bridge owners who are engaged in or contemplating any construction operation on their bridges to ensure that any construction loading and stockpiled raw materials placed on a structure do not overload its members.

For additional information, the advisory referred State agencies to the AASHTO *Standard Specifications for Highway Bridges*, 17th edition, division II, section 8.15, or the AASHTO *Load and Resistance Factor Design Bridge Design Specifications*, 4th edition, section 3.

The referenced AASHTO standard specifications stated, in part:

> loads imposed on existing, new or partially completed portions of structures due to construction operations shall not exceed the load carrying capacity of the structure, or portion of the structure, as determined by the load factor design methods of AASHTO using load group 1B.

---

[16] The Virtis software program is based on the AASHTO *Standard Specifications for Highway Bridges*. A related software program, Opis, uses the AASHTO *Load and Resistance Factor Design Bridge Design Specifications*. Neither Virtis nor Opis models gusset plates.

Mn/DOT subsequently revised section 1513 of its *Standard Specifications for Construction* and added the following paragraph:

> Unless specifically allowed in the Contract, or approved by the Engineer, all construction material and/or equipment which might be temporarily stored or parked on a bridge deck while the bridge is under construction will be limited by this specification. These requirements are intended to limit construction loads to levels commensurate with the typical design live load. The storage of materials and equipment as a whole will be limited to all of the following:
>
> Combinations of vehicles, materials, and other equipment are limited to a maximum weight of 31,702 kg/100 m^2 (65,000 lbs./1000 ft^2).
>
> Material stockpiles (including but not limited to pallets of products, reinforcing bar bundles, aggregate piles) are limited to a maximum weight of 12,200 kg/10 m^2 (25,000 lbs./100 ft^2).
>
> Combinations of vehicles, materials, and other equipment are limited to a maximum weight of 90,700 kg (200,000 lbs.) per span.
>
> The Contractor may submit alternate loadings to the Project Engineer 30 Calendar days prior to placement. Any submittals will require the calculations be certified by a Professional Engineer.

Bridge Load Rating and Posting

In 1970, the AASHTO *Manual for Maintenance Inspection of Bridges* introduced procedural guidance to be used in determining the inventory and operating ratings for bridges. At the same time, the National Bridge Inspection Standards were being developed and implemented through Federal regulations. These standards require that each bridge be load rated to determine its safe load-carrying capacity. These ratings must be performed whenever a significant change occurs that could affect the bridge's load-carrying capacity. For example, a load rating would be required for a bridge undergoing a renovation or rehabilitation that increased the dead load on the structure. A bridge would also need to be load rated if an inspection revealed deterioration that called into question the ability of the bridge to safely continue operating at its previous load rating. There are no requirements for a load rating to occur before a new bridge is opened to traffic.

A bridge's load rating is used by a State to determine whether to approve a request to move loads larger than the established legal loads over the structure. If the requested load exceeds the load rating of a bridge, the load may be redirected to other routes with bridges having sufficient ratings for them to safely carry the permitted load.

Typically, the bridge designer performs the initial load rating for each member of a bridge, with the member with the lowest load rating used to classify the entire structure. Unless inspections reveal deterioration or other conditions that would reduce the capacity of structural members other than those initially identified as having the lowest load capacity, these members continue to be the basis for subsequent load ratings. In the case of the I-35W bridge, the design firm provided Mn/DOT with the capacity for each member in the truss and for both approach spans, as well as influence lines[17] for the main truss members, floor trusses, and the south and north approach girders. No information was found regarding the capacity of the gusset plates. Additionally, no documentation was found to show which member was classified as the critical or controlling member of the bridge until 1995 (as discussed later in this section).

Bridge load ratings must be performed in accordance with guidelines established by AASHTO. This requirement, as well as the directive to the States to perform ratings in accordance with AASHTO (formerly AASHO) guidance, dates to the inception of the National Bridge Inspection Standards program. At the time of the collapse, bridge load rating in the United States was guided by the AASHTO *Manual for Condition Evaluation of Bridges*, 2nd edition (2000), and the *Manual for Condition Evaluation and Load and Resistance Factor Rating (LRFR) of Highway Bridges*, 1st edition (2003). The *Manual for Condition Evaluation of Bridges* states:

> Bridge load rating calculations provide a basis for determining the safe load capacity of a bridge. Load rating requires engineering judgment in determining a rating value that is applicable to maintaining the safe use of the bridge and arriving at posting and permit decisions. Bridge load rating calculations are based on information in the bridge file including the results of a recent inspection. As part of every inspection cycle, bridge load ratings should be reviewed and updated to reflect any relevant changes in condition or dead load noted during the inspection.

Bridges are rated at two levels of stress: inventory and operating. The *inventory* level is equivalent to the design level of stress. A bridge subjected to no more than the inventory stress level can be expected to safely function indefinitely. The *operating* level is the maximum permissible live load[18] stress level to which a structure may be subjected. This rating is used by Mn/DOT and most other State DOTs in evaluating vehicles for overweight permits and in determining whether a bridge should be posted with maximum allowable loads. A bridge that is subjected to the operating stress level for extended periods may be expected to have a reduced service life.

---

[17] An *influence line* for a given function—such as a reaction, axial force, shear force, or bending moment—is a graph that shows the variation of that function due to the application of a unit load at any point on the structure.

[18] *Live load* refers to operational or temporary loads, such as vehicular traffic, impact, wind, water, or earthquake.

Several load rating methods have been used to calculate inventory and operating stress levels. The AASHO *Manual for Maintenance Inspection of Bridges* (1970) used the *allowable stress* method, which compared stresses caused by the actual loadings on a structure to allowable stresses. Later, the AASHTO *Manual for Maintenance Inspection of Bridges* (1978) added the *load factor* method, in which bridge loadings are factored up individually and compared to capacities based on yield stress of the material. The AASHTO *Manual for Condition Evaluation and Load and Resistance Factor Rating (LRFR) of Highway Bridges* (2003) used the more recently developed *load and resistance factor* rating method, which is a reliability-based design methodology in which force effects caused by factored loads are not permitted to exceed the factored resistance of the components. In 2008, AASHTO combined these three load rating methods into a single publication, the *Manual for Bridge Evaluation*. Mn/DOT calculated load ratings in accordance with the AASHTO *Manual for Condition Evaluation of Bridges* (2000), which gives the following general expression in determining the load rating of a bridge:

Load rating factor (RF) = $(C - A_1 DL)/(A_2 LL)$

where:

RF = rating for live-load carrying capacity (rating factor multiplied by rating vehicle in tons yields structure rating)

C = capacity of member

DL and LL = dead load and live load effect on member

A_1 = factor for dead loads

A_2 = factor for live loads.

The AASHTO manual gives specific values for A_1 and A_2 depending on which load rating method (allowable stress or load factor) and which rating level (inventory or operating) are used. The formula above should be applied to all of the critical sections of the bridge.

The *Manual for Condition Evaluation of Bridges*, as did previous AASHTO guidance, considered the capacity of bridge members but did not specifically state that connections (gusset plates) should be evaluated or provide a method for their evaluation. The latest guidance, as contained in the AASHTO *Manual for Bridge Evaluation*, does not provide information on the evaluation of gusset plates.

The load rating factor, which is typically expressed in tons, may be used to determine the rating of the bridge member as follows:

RT = (RF)W

where:

RT = bridge member rating in tons

RF = load rating factor

W = weight (in tons) of nominal truck used in determining live load effect (Minnesota uses the HS20, or 36 tons, nominal truck in determining live load).

Mn/DOT Draft Bridge Rating Manual

Mn/DOT provided the Safety Board with a copy of its draft *LRFD* [load and resistance factor design] *Bridge Design Manual*, dated June 2007. Chapter 15 of the draft manual, titled "Bridge Rating," states, in part:

From the "Introduction" section:

Bridge ratings are administered and performed by the Bridge Rating Unit of the Mn/DOT Bridge Office. Bridge ratings may also be performed by other qualified engineers.

All bridges in Minnesota open to the public, with spans of 10 feet and more are rated. This includes all county and local bridges. However, bridges that carry pedestrians, recreational traffic, or railroad trains need not be rated.

Rating results are kept on file, and key information is entered in the Pontis database. From there annual reports are prepared and sent to the FHWA.

Bridge Ratings are calculated in accordance with the AASHTO *Manual for Condition Evaluation of Bridges* (MCE).

From the "Glossary" section:

Design Load Rating: The AASHTO HS truck and lane loads are used for the live load. The final rating is usually expressed relative to HS20. This is usually calculated at both the inventory and operating levels.

Legal Load Rating: (Sometimes called Posting Rating.) The live load is one or more of the 'legal trucks.' If the RF is less than 1.00 (or another specified amount), the bridge will be posted.

RF: Rating Factor: The result of calculating the rating equation, MCE 6-1a. Generally RF≥1.0 indicates that the member or bridge has sufficient capacity for the equated live load and is acceptable; and RF<1.0 indicates overstress and requires further action. The RF may be converted to a weight by applying the equation, MCE 6-1b. An RF is always associated with a particular live load . . .

From the "General" section:

Bridges are rated at two different stress levels, Inventory level and Operating level. The Operating level is used for load posting and for evaluation of overweight permits.

In almost all cases only the primary load carrying members of the superstructure are rated. Decks or piers may have to be investigated in unusual circumstances such as severe deterioration. Unusually heavy permit loads may also require investigation of the deck and piers.

When rating a bridge, the final overall bridge rating should be the rating of the weakest point of the weakest member within the bridge. This is recorded on the cover sheet of the rating form.

From the "Loads" section:

For steel bridges, account for the extra dead loads such as welds, splices, bolts, connection plates, etc. This generally ranges from 2 percent to 5 percent of the main member weight.

Design ratings are calculated and reported in terms of HS20. Thus with the HS20 truck as the live load in the denominator of the rating equation and if the resulting rating factor is 1.17, the rating would be recorded as HS23.4.

From the "Rating New Bridges" section:

New bridges are to be rated anytime after the plan is completed and before the bridge is opened to traffic. The results are then turned in to the Bridge Management Unit for entering in Pontis.

For Mn/DOT bridges, the records remain inactive until Bridge Management is informed that the bridge has been opened to traffic.

If any changes are made to the bridge during construction that would affect the rating, these changes should be reported to the Bridge Ratings Unit (or the person who did the original rating), and also be recorded on the as built plans. This includes strand pattern changes for prestressed beams. The bridge rating is then recalculated.

From the "Rerating Existing Bridges" section:

A new bridge rating should be calculated whenever a change occurs that would affect the rating. The most commonly encountered types of changes are:

A modification that changes the dead load on the bridge (For example: a deck overlay)

Damage that alters the structural capacity of the bridge (For example: being hit by an oversize load)

Deterioration that alters the structural capacity of the bridge (For example: rust, corrosion or rot). Scheduled inspections are usually the source of this information.

Settlement or movement of a pier or abutment.

Repairs or remodeling.

A change in the AASHTO Rating Specification.

An upgrading of the rating software.

A change in laws regulating truck weights.

A new rating should be completed, signed, dated, and filed, as outlined in the Forms and Documentation Section of this chapter. This most recent rating then supersedes any and all preceding ratings.

The 2007 Mn/DOT *LRFD Bridge Design Manual* includes provisions for conducting a load rating on a new bridge before it is opened. According to Mn/DOT, it had been informally conducting such load ratings for the past 10 years. The Mn/DOT directive to perform load ratings on new bridges differs from AASHTO guidance, including the recently published *Manual for Bridge Evaluation*. The AASHTO guidance requires that load ratings be performed whenever a significant change occurs that could affect a bridge's load-carrying capacity but has no provision for load rating a new bridge before it is put into service.

According to Mn/DOT, the decision to load rate new bridges was based in part on the existing reporting requirements within the National Bridge Inspection Standards. As part of its report for the National Bridge Inventory, the FHWA requires that each State annually submit the inventory and operating ratings for all bridges.

Load Ratings for I-35W Bridge

Table 3 summarizes the history of superstructure inventory and operating ratings for the I-35W bridge from 1983 until its collapse in 2007.

Table 3. Summary of load ratings for I-35W bridge, 1983–2007.

Years[A]	Inventory rating (U.S. tons)[B]	Operating rating (U.S. tons)
1983–1995	26.8 (HS14.9)[C]	53.6 (HS29.8)
1996–1998	35.7 (HS19.8)	58.5 (HS32.5)
1999–2001	35.7 (HS19.8)	59.0 (HS32.8)
2002–2007	36.0 (HS20)	59.4 (HS33)

[A]Although data exist from as early as 1979, these data were originally maintained in file formats that did not allow for simple conversion into current definitions. Starting with 1983, the FHWA was able to provide data it believed were accurate and consistent with current record-keeping.

[B]The inventory and operating ratings shown on the National Bridge Inventory are reported in metric tons and have been converted to U.S. tons.

[C]The inventory and operating ratings can also be expressed in terms of an HS20 vehicle, or 36 tons, as a nominal truck in determining live load. The conversion from U.S. tons to an HS20 vehicle can be computed as follows: (26.8 tons / 36 tons) x HS20 = HS14.9.

Bridge Posting

According to the National Bridge Inspection Standards, the posting of a maximum weight limit sign is required if the maximum vehicle weight that State regulation allows (in Minnesota, 80,000 pounds) on that highway exceeds the bridge's maximum weight limit as determined by the operating rating, as specified below:

> If it is determined under this rating procedure that the maximum legal load under State law exceeds the load permitted under the Operating Rating, the bridge must be posted in conformity with the AASHTO Manual or in accordance with State law.

Mn/DOT's *Bridge Rating Manual* stated that when the operating load rating factor (RF) is less than 1, the bridge will be posted. From the data in table 3, the operating rating for the I-35W bridge varied from 53.6–59.4 tons from 1983–2007. Based on the formula (discussed above) RT = (RF)W, the load rating factor for the I-35W bridge was greater than 1, as shown below:

RF = RT / W = (53.6 tons) / (36 tons) = 1.49

The posting of a maximum weight limit sign on the I-35W bridge was not required because the operating load rating factor (RF) for all legal trucks was never below 1.

Bridge Rating and Load Posting Report, 1979

A September 17, 1979, bridge rating and load posting report for the I-35W bridge indicated an inventory rating of HS15.9 (or 28.6 tons) and an operating rating of HS30.6 (or 55.1 tons). Posting of the bridge was not required. This report was certified by a registered professional engineer in the State of Minnesota. The only documentation available to Safety Board investigators was the one-page load rating summary report. Because this report listed both the inventory and operating ratings, it was assumed that Mn/DOT had followed its policy of calculating the bridge load rating in accordance with AASHTO guidelines. Additionally, this load rating post-dated the 1977 deck overlay project, which had substantially increased the dead load of the structure. Investigators were unable to determine if the load rating had occurred before the 1977 project and was officially documented in 1979 or if the rating resulted from the project. A search of bridge inspection reports did not reveal any information of a condition or event that would have required that a load rating be performed in 1979, leading investigators to believe that the rating had been performed in conjunction with the 1977 deck overlay project.

Bridge Rating and Load Posting Report, 1995

A December 14, 1995, bridge rating and load posting report for the I-35W bridge indicated an inventory rating of HS20 (or 36 tons) and an operating rating of HS33 (or 59.4 tons). This report was also certified by a registered professional engineer in the State of Minnesota. A review of both the load rating report and supporting attachments indicated that this load rating was performed in accordance with National Bridge Inspection Standards requirements and associated AASHTO guidance. As such, the load rating would not have included an evaluation of gusset plate capacity. Posting of the bridge was not required. The controlling section of the bridge was identified in the south approach spans (spans 1–5). Mn/DOT officials stated the following regarding the controlling section of the bridge:

> This controlling rating is from the SB roadway, beam line G13, at midspan in the fourth of the five continuous spans (107.25 ft). The limit state is tension stress in the bottom flange due to bending moment. The rating program used here, BARS, is a line analysis program. Member G13 would have been selected as a good representative because of its longer length in span one and it is the first interior beam.

Mn/DOT officials attributed the fact that the load ratings in 1995 were higher than those in 1979 to the fact that the earlier ratings were calculated using the allowable stress method. They stated that the load factor method, which was used for the 1995 rating, "typically yields higher rating numbers." Mn/DOT was not able to determine from the 1979 rating sheet which portion of the bridge controlled the rating. Also, the rating sheet in the bridge management file did not include a BARS computer printout.

The 1995 load rating was based on a BARS analysis that calculated the inventory and operating ratings for three critical sections of the bridge: S01, representing the south approach spans (spans 1–5); S02, representing the north approach spans (spans 9–11); and S03, representing the far north approach spans (spans 12–14). (See figure 18.) The dead load used in the rating analysis was 358 pounds per linear foot for critical sections S01 and S02 and 47 pounds per linear foot for critical section S03.

On August 18, 1997, the same Mn/DOT engineer who performed the 1995 load rating analysis calculated a new dead load for the I-35W bridge to reflect the weight of the new median barrier and outside traffic railings, which were to be constructed in 1998. The engineer's calculations showed a new dead load of 487 pounds per linear foot for critical sections S01 and S02. A new rating analysis was done for the truss stringers that had not been included in the 1995 analysis. This calculation showed a dead load of 487 pounds per linear foot for critical sections S04 and S05. The previously calculated 47 pounds per linear foot for critical section S03 remained unchanged because it was at the end of the bridge and beyond the limits of the area of improvement.

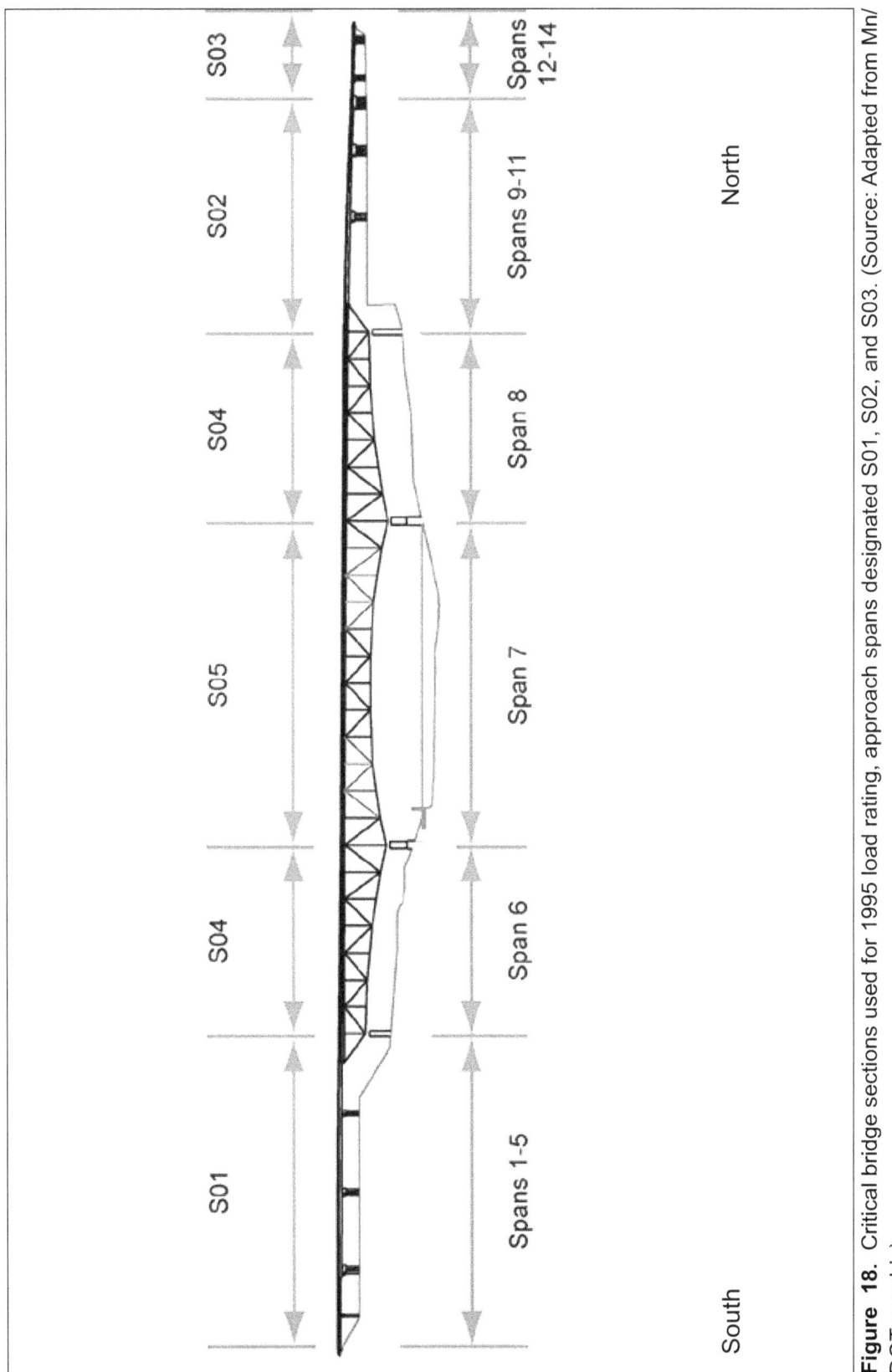

Figure 18. Critical bridge sections used for 1995 load rating, approach spans designated S01, S02, and S03. (Source: Adapted from Mn/DOT graphic)

Table 4 compares the inventory and operating load ratings as shown on the December 1995 bridge rating and load posting report with what was calculated in the December 11, 1995, and August 18, 1997, BARS computer analyses.

Table 4. Comparison of I-35W bridge load ratings with findings from BARS analyses.[A]

Critical section	December 1995 bridge rating and load posting report	December 11, 1995, BARS computer printout		August 18, 1997, BARS computer printout	
		Rating	Dead load (lb/ft)	Rating	Dead load (lb/ft)
S01	HS20 (IR) HS33 (OR)	HS19.8 (IR) HS32.9 (OR)	358	HS18.9 (IR) HS31.5 (OR)	487
S02	HS27 (IR) HS45.5 (OR)	HS27.3 (IR) HS45.5 (OR)	358	HS26.7(IR) HS44.5 (OR)	487
S03	HS20.2 (IR) HS33.6 (OR)	HS20.2 (IR) HS33.6 (OR)	47	HS20.2 (IR) HS33.6 (OR)	47
S04, S05[B]	HS22.8 (IR) HS38.1 (OR)	Not calculated	--	HS22.8 (IR) HS38.1 (OR)	487

[A]Inventory rating (IR), operating rating (OR).

[B]The inventory and operating ratings for critical sections S04 and S05 are the same because they both consist of truss stringers. The inventory and operating ratings shown are for S04.

The inventory and operating load ratings were lower in the latter analysis for two critical sections as a result of increased dead load. These lower ratings were not reflected on the bridge rating and load posting reports. When asked by the Safety Board why the new bridge load ratings had not been officially documented, Mn/DOT officials responded:

> It appears that a new rating was computed with BARS in August 1997, before the construction work was done on the bridge. The construction contract was bid on March 27, 1998, with work performed during the 1998 construction season. Apparently the follow up to officially document and record the rating did not occur after construction was completed.

FHWA Assessment of Load Rating Records for I-35W Bridge

The FHWA Turner-Fairbank Highway Research Center, on June 30, 2008, released *Assessment of the Load Rating Records for Minnesota Bridge No. 9340 (I-35W Over the Mississippi River)*. The findings of the report include the following:

> No information on load rating of the truss portion of the structure was found in the documentation supplied for any of the load ratings conducted. The load rating file should include an analysis supporting the current load rating

for the entire bridge, including the deck truss, either from an initial analysis concluding that the truss was not critical to future ratings or from a rating prompted by a change in conditions or deterioration. The influence lines that were included for all truss members in the original design documents may have been used initially to verify that the rating was controlled by the deck stringer system; however, there is nothing within the documentation provided to support this assumption. A re-rating was performed on the approach spans in 1995, and again in 1997; however, no information was included pertaining to a re-rating of the truss structure. The re-rating in 1997 was warranted due to an increase in dead load resulting from the change in bridge barrier type.

The only document retained from the 1979 load rating was the Rating and Load Posting Report Sheet. The Report Sheet indicates a reduction in capacity of approximately 20 percent from design values. While no supporting documentation was reviewed, it can be inferred from calculation and other information that the reduction in rating was due to the added weight of the 1977 bridge overlay.

The retained records for the load ratings conducted in 1995 and in 1997 on the approach spans are incomplete. These ratings were conducted on the interior G13 girder. It is unclear if this is the controlling girder line and unknown whether the engineer considered other girder lines.

The dead load calculations retained from the 1997 rating contained minor errors. The height for the exterior barrier installed in the 1990's should have been 2'-8" instead of 2'-0" and the width should have been 10 inches instead of 9 inches. Also, diaphragms, lamp posts, and existing metal posts were not included in the dead load calculations. Although the overall significance of these items may be minimal, a load rating analysis should accurately account for all existing dead load conditions applied to the structure and include a narrative describing what assumptions were made in determining the applied dead load.

The inspection reports indicate that several of the exterior approach span girders, primary truss and floor truss members, and primary truss connections exhibited some section loss due to corrosion that was not addressed in either a narrative summary or in re-rating calculations. A load rating analysis should take into consideration the loss of capacity resulting from deterioration of all load carrying structural elements or the file should include a discussion detailing the reasons why the deterioration was considered negligible.

The most recent "Load Rating Summary" sheet is not correct. In the 1997 calculations provided, the girders in the south approach governed with an HS18.9 Inventory Rating and a HS31.5 Operating Rating (ratings in this document are reported in Customary U.S. Units and are based on an HS20 live load rating vehicle and the load factor rating method). The controlling ratings shown on the most recent Load Rating Summary sheet are HS20.0 for Inventory and HS33.0 for Operating. This error seems to have resulted from not appropriately updating the information included on the 1995 load rating summary sheets. It appears that the stringer calculations conducted in 1997 were simply appended to the 1995 Load Rating Summary sheet and no new summary sheet was generated despite the increase in bridge rail dead load which resulted in about a 5 percent reduction in load carrying capacity.

According to the 1979 Load Rating Summary sheet, the Inventory Rating was HS15.9 and the Operating Rating was HS30.6. According to the 1995 Load Rating Summary Sheet, the Inventory Rating was HS20 and the Operating Rating was HS33. While it is most likely that the variation in ratings between 1979 and 1995 was the result of transitioning from the ASR method to the LFR method that took place during that time period, no documentation was found that provided that explanation.

It is important to note that despite the omissions and inconsistencies of the documentation, the results for all of the ratings conducted indicate that the I-35W Bridge was capable of safely carrying the live load for which it was designed.

Federal Bridge Inspection Requirements

Under the National Bridge Inspection Standards (23 CFR Part 650), each State transportation department must, at regular intervals not to exceed 24 months,

> inspect, or cause to be inspected, all highway bridges located on public roads that are fully or partially located within the State's boundaries, except for bridges that are owned by Federal agencies.

Federal regulations define eight types of bridge inspections, summarized as follows:

> **Damage inspection:** An unscheduled inspection to assess structural damage resulting from environmental factors or human actions.

> **Fracture-critical member inspection:** A hands-on inspection of fracture-critical members or member components that may include visual and other nondestructive evaluation. Federal regulation requires that these inspections be conducted at 24-month intervals.

Hands-on inspection: Inspection within arms length of the component using visual techniques that may be supplemented by nondestructive testing.

In-depth inspection: A close-up inspection of one or more members above or below the water level to identify any deficiencies not readily detectable using routine inspection procedures. Hands-on inspection may be necessary at some locations.

Initial inspection: First inspection of a bridge as it becomes a part of the bridge inventory to provide all structure inventory, appraisal, and other relevant data and to determine baseline structural conditions.

Routine inspection: Regularly scheduled inspection consisting of observations or measurements needed to (1) determine the physical and functional condition of the bridge, (2) identify any changes from initial or previously recorded conditions, and (3) ensure that the structure continues to satisfy present service requirements. These inspections are required to be performed at 24-month intervals.

Special inspection: An inspection scheduled at the discretion of the bridge owner, used to monitor a particular known or suspected deficiency.

Underwater inspection: Inspection of the underwater portion of a bridge substructure and the surrounding channel that cannot be inspected visually at low water by wading or probing. These inspections generally require diving or other appropriate techniques and are required to be performed at 60-month intervals.

State Inspections and Inspection Reporting

Mn/DOT performed the first inspection of the I-35W bridge in 1971, shortly after implementation of the National Bridge Inspection Standards; and records indicated that the bridge continued to be inspected annually. In accordance with Federal requirements, Mn/DOT also conducted regularly scheduled in-depth fracture-critical inspections of the I-35W bridge. Beginning in 1994, Mn/DOT began conducting these inspections annually. Both an in-depth fracture-critical and a routine inspection of the bridge were completed in June 2006. The findings from these and earlier inspections of the I-35W bridge are discussed later in this report.

During the 40-year life of the I-35W bridge, Mn/DOT bridge inspectors used a variety of formats and forms to record the findings of their inspections. The inspection reports used from 1971–1973 listed about 22 components of the substructure, superstructure, deck, channel protection, culverts, retaining wall, approaches, and signs that were to be inspected and rated. The inspection report forms used from 1974–1987 typically consisted of 24 rating elements. The forms used from 1988–1993 consisted of 35 elements. Mn/DOT's current inspection report covers about 150 elements.

At the time of the accident, Mn/DOT's reports were produced by a software-based bridge management system called Pontis, which Mn/DOT has used since 1994 to document its inspections and since 2000 to create the files for submittal to the FHWA for calculation of bridge sufficiency and status ratings. Pontis was developed for the FHWA and is licensed to State DOTs and other agencies by AASHTO. Findings from bridge inspections are fed into the Pontis system, which uses internal models and algorithms to analyze the data and make predictions to help transportation officials plan future bridge inspections or maintenance. The system is currently used by about 45 States. Mn/DOT officials told the Safety Board that the agency uses Pontis inspection data in combination with detailed spreadsheets to develop a 20-year plan to help identify bridges that need rehabilitation or replacement due to condition, age, and traffic volume.

The Pontis system provides a detailed condition rating of a bridge by dividing the structure into separate elements that inspectors rate individually based on the severity and extent of deterioration. An "element" refers to structural members (such as beams, pier columns, or decks) or any other components (for example, railings, expansion joints, or approach panels) commonly found on a bridge. This rating system was developed by AASHTO and is outlined in the *AASHTO Guide for Commonly Recognized (CoRe) Structural Elements.*

The AASHTO CoRe element descriptions were developed by highway engineers representing six State highway departments and the FHWA. AASHTO states that the element descriptions are not unique to Pontis; they are intended for use as a basis for data collection in any bridge management system, which should facilitate data sharing among States. Among the "Non-CoRe Elements" listed in the guide are gusset plates. The manual states:

> Connectors for steel elements (splice plates, etc.) are not identified as CoRe elements because of the inability to accurately model the deterioration rate of missing bolts, etc. These phenomena are handled with Smart Flags.

The guide describes a "Smart Flag" as follows:

> A Smart Flag is similar to an element in that it will have multiple stages of deterioration. However, a Smart Flag does not have feasible actions. . . The Smart Flags will allow States to track distress conditions in elements that do not follow the same deterioration or do not have the same units of measure as the distress described in the CoRe element.

The 150 elements in the Mn/DOT bridge inspection manual included the AASHTO CoRe elements as well as elements added by Mn/DOT to better represent the bridge types and components found in Minnesota. Mn/DOT did not include gusset plates as a separate inspection element.

Inspection Results and Condition Ratings for I-35W Bridge

Under the National Bridge Inspection Standards, bridges are inspected and rated as to the condition of their deck, superstructure, and substructure. Based on these ratings and other factors, each bridge is assigned a sufficiency rating and a status.

The *condition* of a bridge deck, superstructure, and substructure, as determined through the required inspections, is indicated by a numerical rating as follows:

9 – Excellent condition

8 – Very good condition

7 – Good condition

6 – Satisfactory condition

5 – Fair condition

4 – Poor condition

3 – Serious condition

2 – Critical condition

1 – "Imminent" failure condition

0 – Failed condition.

The *sufficiency rating* of a bridge is a computed numerical value that is used to determine the eligibility of a bridge for Federal funding. This rating formula returns a value from 0–100 and includes factors for structural condition, bridge geometry, and traffic considerations.[19] A bridge with a sufficiency rating of 80 or less is eligible for Federal bridge rehabilitation funding. A bridge with a sufficiency rating of less than 50 is eligible for Federal bridge replacement funding.

Based on the bridge's condition, a *status* is assigned. The status is used to determine eligibility for Federal bridge replacement and rehabilitation funding. Current FHWA status ratings are: Not Deficient, Structurally Deficient, and Functionally Obsolete.

---

[19] The sufficiency rating formula is contained in the December 1995 edition of the FHWA's *Recording and Coding Guide for the Structure Inventory and Appraisal of the Nation's Bridges* (Report No. FHWA-PD-96-001).

A bridge is rated *Structurally Deficient* if it has a general condition rating for the deck, superstructure, substructure, or culvert of 4 (poor condition) or less, or if the road approaches to the bridge regularly overtop due to flooding. Examples of poor condition include corrosion that has caused significant section loss of steel support members, movement of substructures, or advanced cracking and deterioration in concrete bridge decks. For bridge owners, this rating is a reminder that the bridge may need further analysis, which may result in load posting, maintenance, rehabilitation, replacement, or closure.

According to the FHWA, a status of *Structurally Deficient* does not indicate that a bridge is unsafe but only that the structure is in need of maintenance, repair, or eventual rehabilitation. If required to remain open to traffic, a *Structurally Deficient* bridge can be posted to restrict the gross weight of vehicles permitted to use it. If unsafe conditions are identified during a physical inspection, the bridge will be closed. According to FHWA data (December 2007), about 72,500, or 12 percent, of the 600,000 bridges in the National Bridge Inventory are currently rated *Structurally Deficient*. Bridges so rated were located in every State and U.S. territory, with no State having fewer than 20 and three States having more than 5,000 each.

Of the 465 steel deck truss bridges in the inventory, 145, or 31 percent, are currently rated *Structurally Deficient*. The four States with the largest number of steel deck truss bridges so rated are California (22 of a total of 50 steel truss bridges rated *Structurally Deficient*), Pennsylvania (16 of 48), Oregon (12 of 37), and Iowa (8 of 9). One of Minnesota's four remaining steel deck truss bridges is rated *Structurally Deficient* and is currently being replaced.

A *Functionally Obsolete* bridge is one that was built to standards that do not meet the current minimum Federal clearance requirements for new bridges. This status rating may apply to bridges that have substandard geometric features, such as narrow lanes, narrow shoulders, poor approach alignment, or inadequate vertical under-clearance. A rating of *Functionally Obsolete* reflects evolving design standards and does not indicate that the structure is unsound. The classification *Functionally Obsolete* is also a term used as a priority status for Federal bridge replacement and rehabilitation funding eligibility. According to FHWA data (December 2007), about 79,800, or 13 percent, of the bridges in the National Bridge Inventory are currently rated *Functionally Obsolete*.

The FHWA provided the Safety Board with ratings for the I-35W bridge for the years 1983–2007. These ratings reflected the bridge deck condition, superstructure condition, substructure condition (except for 1999, when no rating was submitted by Mn/DOT), sufficiency rating, and status, as determined through the required physical inspections of the bridge. (See table 5.)

Table 5. Condition ratings, sufficiency ratings, and status for I-35W bridge, 1983–2007.

Year[A]	Deck condition rating	Superstructure condition rating	Substructure condition rating	Sufficiency rating	Status
1983	6	7	6	80.1	Not deficient
1984	6	7	6	80.1	Not deficient
1985	6	7	6	80.1	Not deficient
1986	6	7	6	79.6	Not deficient
1987	6	7	6	79.8	Not deficient
1988	6	7	6	79.8	Not deficient
1989	6	8[B]	6	75.5	Not deficient
1990	6	7	6	75.5	Not deficient
1991	6	4	6	46.5	Structurally deficient
1992	6	4	6	46.5	Structurally deficient
1993	6	4	6	46.5	Structurally deficient
1994	6	4	6	46.5	Structurally deficient
1995	6	4	6	46.5	Structurally deficient
1996	6	4	6	49	Structurally deficient
1997	6	4	6	49	Structurally deficient
1998	6	4	6	49	Structurally deficient
1999	N/A	N/A	N/A	76	Not deficient[C]
2000	5	4	6	48	Structurally deficient
2001	5	4	6	48	Structurally deficient
2002	5	4	6	50	Structurally deficient
2003	5	4	6	50	Structurally deficient
2004	5	4	6	50	Structurally deficient
2005	5	4	6	50	Structurally deficient
2006	5	4	6	50	Structurally deficient
2007	5	4	6	50	Structurally deficient

[A]Although data exist from as early as 1979, these data were originally maintained in file formats that did not allow for simple conversion into current definitions. Starting with 1983, the FHWA was able to provide data it believed were accurate and consistent with current record-keeping.

[B]Mn/DOT officials attributed this apparent improvement in rating from the previous year to an error in the data file submitted to the FHWA.

[C]According to Mn/DOT, the bridge inspection reporting software that was being used at the time (Brinfo) did not submit the correct 1999 inspection data to the FHWA, and the FHWA system used default values to calculate the sufficiency rating and status. In 2000, Mn/DOT began using the Pontis system to format and submit the data.

The I-35W bridge had been classified *Structurally Deficient* since 1991, when the superstructure received its first condition rating of 4 (poor condition). The bridge superstructure had a recorded condition rating of 4 on each of the National Bridge Inventory forms from 1991–2007, including 1999 when the condition rating was not properly submitted to the FHWA.

The National Bridge Inventory condition rating guidelines were included in Mn/DOT's *Bridge Inspection Manual* dated May 2007. The guidelines contained a superstructure condition description for a rating of 4 as follows:

> **Poor Condition:** Superstructure has advanced deterioration. Members may be significantly bent or misaligned. Connection failure may be imminent. Bearings may be severely restricted.
>
> Steel: significant section loss in critical stress areas. Un-arrested fatigue cracks exist that may likely propagate into critical stress areas.
>
> Concrete: advanced scaling, cracking, or spalling (significant structural cracks may be present–exposed reinforcement may have significant section loss).
>
> Timber: advanced splitting (extensive decay or significant crushing).
>
> Masonry: advanced weathering or cracking (joints may have separation or offset).

Safety Board investigators reviewed Mn/DOT inspection reports for the I-35W bridge dating from 1971–2006. Until the 1991 inspection, the bridge superstructure was assigned a rating of at least 7 (good condition) because all of the superstructure elements (trusses, girders, floor beams, stringers or beams, bearing devices, arches, fascia beams, diaphragms, and spandrel columns) received a rating of 7 or above. The rating was lowered in 1991 because of the condition of bearing devices. The 1993 report rated the superstructure as 4 (poor condition) based on a rating of 4 for one element (bearing devices in the south approach spans). The remaining elements (trusses, girders, floor beams, stringers or beams, arches, fascia beams, diaphragms, and spandrel columns) received a rating of 7 or above. Comments in the 1993 report pertaining to the bearing devices on the south approach spans included the following:

> Last four bearing plates south abutment west side are quite rusty.
>
> Bearings on Span #1 cantilever section are closed tight at 60 degrees F.
>
> Bearing pins on truss bearing assemblies at ends of truss should be replaced with slightly longer bolts to allow for thermal thrust (on [*sic*] even expansion–due to temperature differences between girders and truss components).

The last regularly scheduled State inspection of the I-35W bridge prior to the collapse was in June 2006. Mn/DOT issued the results of this fracture-critical inspection in its *Fracture Critical Bridge Inspection In-Depth Report, Bridge #9340 (Squirt Bridge), I-35W Over the Mississippi River at Minneapolis, MN.* Inspectors assigned a condition rating of 4 (poor) to the bridge superstructure based, in part, on the following (abridged) findings:

Paint System: . . . the overall paint system is approximately 15 percent unsound. The truss members have surface rust corrosion and pack rust[20] at the floorbeam & sway frame connections, and there is paint failure & surface rust corrosion in scattered locations. The floorbeam trusses & stringer ends have surface rust corrosion at the stringer expansion joints. Some of the areas re-painted in 1999 have severe section loss. This includes the sections of the floorbeam trusses & sway bracing located below the median, and the truss end floor beams & "crossbeams", located below the open finger joints.

Main Truss Members: . . . The truss members have numerous poor weld details. . . . The truss members have surface rust corrosion at the floor beam and sway frame connections. Pack rust is forming between the connection plates. There is paint failure, surface rust, and section loss, flaking rust in scattered locations.

Floor Beam Trusses: . . . The floorbeam truss members have numerous poor welding details, including plug welded web reinforcement plates, and tack welds & welded connection plates located in tension zones. Some of the top chord splices are offset vertically, up to 1/2" – from original construction. The splice plates are bent. The floorbeam trusses below stringer joints have section loss, severe flaking rust. There is pack rust and surface pitting at the main truss connections.

Stringers: . . . The stringer ends have surface rust corrosion at the expansion joints. . . . The bolted connections to the floorbeam trusses are 'working' and some bolts are loose or missing. [2006] [*sic*] Fascia stringers have minor section loss, with moderate flaking rust along the bottom flange.

Truss Bearing Assemblies: . . . The truss bearings have section loss, flaking & surface rust; moderate corrosion, the bearings at piers #5 & 8 are functioning properly. They are checked during each annual [routine] inspection. The bearings at pier #6 show no obvious signs of movement, difficult to reach with snooper.[21]

End Floor Beams: . . . The sides facing the open finger joints have extensive section loss with surface pitting at the base of the web, and holes in the base of the vertical stiffeners.

Crossbeams & Rocker Bearings: . . . The faces exposed to the finger joints have extensive surface pitting with some areas of severe section loss with holes at the base of stiffeners. The rocker bearings are measured & checked

---

[20] *Pack rust* is a thick buildup of corrosion product that tends to develop between the surfaces of closely joined metal objects and can force the joint apart.

[21] A *snooper* is an inspection bucket or platform at the end of a long articulating boom (usually mounted to a truck) that provides access to the undersides of bridges.

for movement during each annual [routine] inspection. All four bearings appear to be functioning. They show obvious signs of movement.

Steel Multi-Beam Approach Spans (spans #1–5 & #9–11): . . . In span #2, multi-beam approach span, there is a cantilever expansion hinge with sliding plate bearings. The joint is closed beyond tolerable limits, possibly due to substructure movement & pavement thrust and is no longer functioning. Some beam-ends are contacting, and some bearing plates have tipped, preventing the joint from reopening. The hinge area, with open finger joint above, was re-painted in 1999. The beam-ends have section loss, moderate surface pitting.

As with previous fracture-critical inspection reports for the I-35W bridge, the 2006 report recommended:

> Due to the 'Fracture Critical' configuration of the main river spans and the problematic 'crossbeam' details, and fatigue cracking in the approach spans, eventual replacement of the entire structure would be preferable.

The report went on to recommend that, if replacement were to be significantly delayed, the bridge should be redecked and its superstructure repainted. Regarding the L11 node gusset plates, the report included a photograph of the inside gusset plate at the L11E node and noted, "Section loss: at gusset plate bottom chord. [2004] [*sic*] Pitting: inside gusset plate connection at L11 toward L10."

The sufficiency rating of the I-35W bridge dropped from 75.5 in 1990 to 46.5 in 1991, and it remained at or below 50 (except for 1999 when data were not available) until the collapse. During the periods when the sufficiency rating was below 50, the bridge was eligible for Federal rehabilitation or replacement funding.

Rust and Corrosion

The 1993 inspection report also mentioned a loss of metal in at least one gusset plate due to corrosion. Specifically, inspectors reported that the inside gusset plate at the L11E node "has loss of section 18" long and up to 3/16" deep (original thickness = 1/2")." The report also stated that at the L13 node east, the "lower horiz. brace between the trusses has 3/16" section loss at riveted angle." Beginning in 1994, Mn/DOT began preparing reports for both routine and fracture-critical inspections. All subsequent reports on gusset plate condition were recorded in the fracture-critical reports rather than in the routine inspection reports. In most cases, the reports repeated condition comments from previous inspections and did not quantify the amount of rust or corrosion. Mn/DOT representatives stated that past inspection reports are an integral part of the State's bridge inspection program and that its bridge inspectors use these reports as a checklist and only make additional notations when changes in condition are noted.

Mn/DOT Underwater Bridge Inspection

In 2004, Mn/DOT contracted with Ayres Associates of Eau Claire, Wisconsin, to inspect the underwater substructure of the I-35W bridge. That inspection was performed on December 8, 2004. The report of the inspection stated that

> The concrete surfaces below the water are in good condition.
>
> Minor scaling was found above the water, but not of the quantity or depth as noted in a previous report. The total area was 2.0 feet square and 1/4-inch deep penetration.
>
> No other significant changes in the structure or channel condition have occurred since the last inspection.

The report concluded that no corrective actions regarding the underwater substructure were needed.

Washout Hole Underneath Bridge

In December 2006, Mn/DOT inspectors noted a 4- by 6- by 2-foot-deep hole in the ground below the bridge between piers 4 and 5, under a south approach span. This hole, originally thought to be a sinkhole, was in the edge of the pavement of a roadway that crossed underneath the bridge and abutted an angled concrete drainage pad. After further investigation, Mn/DOT maintenance workers determined that the hole was not a sinkhole but a washout hole caused when water draining from the bridge fell onto the concrete drainage pad and began to work its way into cracks in the concrete. The water began to wash away the earth beneath the concrete pad, eventually pulling some of the dirt from underneath the blacktop of the roadway. The pavement over the void then collapsed, causing the hole.

Mn/DOT maintenance crews made temporary repairs to the washout hole in January 2007. They returned in July to make permanent repairs using a number of loads of concrete and rock. The last time the crews worked on the washout hole was July 25, a week before the collapse.

Other Repairs

In 1986, Mn/DOT inspectors found some cracking of the south side approach span cross girder near the U0E node of the main truss. Resistance to movement of the bearings appeared to have caused significant out-of-plane forces and associated distortion on the cross girder, leading to the formation of cracks. A portion of the bridge had to be jacked up so that the cross girder could be retrofit by drilling holes at the tips of the cracks and adding struts from the reinforcing stiffeners back to the girders. Similar cracking was found in 1997 on the north cross girder near the U0' nodes, and a similar retrofit was performed.

Fatigue Cracking in Minnesota Bridges

According to the Minnesota State bridge engineer, Mn/DOT had for a number of years been concerned about potential fatigue cracking in the State's bridges. It was particularly concerned about non-load-path-redundant bridges, where the failure of a tension member due to fatigue cracking could lead to a catastrophic failure. Because of a general lack of understanding in the 1950s and 1960s of the effects of cyclical loading on steel bridges, many bridges built during that period had poor fatigue-resistant details. Since about 1975, Mn/DOT has conducted fatigue studies on seven State bridges, including the I-35W bridge, where inspections had found evidence of fatigue cracking. The findings of these studies for bridges other than the I-35W bridge (which is discussed in the next section of this report) are summarized below.[22]

Lafayette Bridge

The Lafayette bridge, built in 1968, carries U.S. Highway 52 over the Mississippi River in St. Paul, Minnesota. The main span of the bridge is a fracture-critical two-girder system with floor beams and stringers. In 1975, inspectors found a crack in a primary girder that had developed in a lateral gusset plate web gap and extended through the bottom flange and through about 75 percent of the height of the web. Mn/DOT spliced the girder to repair the crack and performed a retrofit to prevent cracking at similar details elsewhere on the bridge. In 2006, TKDA Consultants, Inc., of St. Paul, conducted a vulnerability assessment of the bridge with regard to fatigue cracking. The bridge is scheduled to be replaced in 2010.

Dresbach Bridge

The Dresbach bridge, built in 1967, carries Interstate 90 over the Mississippi River in the southeastern part of the State. The bridge is a fracture-critical two-girder system with floor beams and stringers. A 1975 inspection found two vertical fatigue cracks in the structure, one of which was 18 inches long. Additional cracks were found in 1987, 1993, 1996, and 1998. Based on a mitigation approach developed by a consultant from Lehigh University, Mn/DOT directed that 1,600 holes be drilled in horizontal lateral bracing plates where welds intersected. A 2005 fatigue study of the bridge by Parsons Brinckerhoff identified numerous other locations with fatigue-sensitive details. The bridge is scheduled to be replaced in 2012.

---

[22] States other than Minnesota have also experienced problems associated with fatigue cracking in bridges. In response, AASHTO, in 1990, published *Guide Specifications for Fatigue Evaluation of Existing Steel Bridges* to assist States in conducting bridge fatigue analyses. Alternative methods for conducting these analyses are contained in chapter 7 ("Fatigue Evaluation of Steel Bridges") of AASHTO's 2003 *Manual for Condition Evaluation and Load and Resistance Factor Rating (LRFR) of Highway Bridges*.

Dartmouth Bridge

The Dartmouth bridge was built in 1964 to carry Interstate 94 over the Mississippi River in Minneapolis. The original superstructure was a steel box that was studied by Parsons Brinckerhoff in 1987 and found to have limited remaining service life. The superstructure was replaced in 1994, primarily due to concerns about potential fatigue.

Lexington Bridge

The Lexington bridge was built in 1963 to carry Interstate 94 over the Mississippi River in St. Paul. After inspections in 1995 found fatigue weld cracks in some support members, Mn/DOT contracted with Lehigh University consultants to develop corrective measures. In late 1995, the diaphragm stiffeners were cut back, and 2-inch-diameter arrester holes were drilled in the girder webs at the ends of stiffener welds. Within a few months, the fatigue cracks reappeared, and more corrective actions were taken. Also, strain gauges were installed, and the repairs were monitored. The cracking was caused by out-of-plane distortion from the floor beam stiffener-to-web weld and was similar to the fatigue cracking found in the I-35W bridge (discussed below). The Lexington bridge was replaced in 2002.

Hastings Bridge

The Hastings bridge, a fracture-critical continuous steel arch truss bridge, was built in 1950 to carry U.S. Highway 61 over the Mississippi River in Hastings, Minnesota. A 1997 inspection found a crack in a tension tie member of the truss arch. The crack was determined to be a brittle fracture and was attributed to an isolated material problem. The member was retrofitted with large cover plates. In 1998, another crack was found in the same member. The crack initiated at a weld at the edge of the floor beam connection gusset plate and had severed one of the web plates. This detail was also retrofitted with cover plates, and the bridge has since been restricted to legal (nonpermitted) loads to limit stress on the members. A project involving painting the bridge and stiffening some gusset plate edges is underway. The bridge is scheduled to be replaced in 2010.

Oar Dock Bridge

The Oar Dock bridge is a multiple-span continuous steel beam superstructure bridge that carries Interstate 35 over railroad tracks in Duluth, Minnesota. Because of the configuration of the tracks, a fracture-critical pier cap was used in the original design to span between columns. When cracks were discovered in the pier cap in 2004, Mn/DOT commissioned a University of Minnesota study that recommended drilling crack-arresting holes. These sites were also specifically monitored during subsequent inspections.

Preexisting Fatigue Cracking in I-35W Bridge

In October 1998, Mn/DOT bridge inspectors found 12 fatigue cracks in 8 of the 48-inch-deep welded approach span girders at the north end of the bridge. The largest of these was an inverted-U-shaped crack, more than 50 inches long, in the web of the third girder from the east, about 20 feet south of pier 9.[23] This crack was in the area where the diaphragm stiffener was welded to the web. The 11 smaller cracks were also near pier 9 and involved 6 additional girders. Each of these cracks occurred at the lower edge of the weld attaching the top flange to the web; each crack penetrated the base metal of the girder and was found at the site of a diaphragm stiffener.

To prevent propagation of the largest crack, bridge maintenance workers initially drilled 2-inch-diameter holes at each end of the crack. After subsequent inspection of the areas of fatigue cracking and consultation with State and University of Minnesota engineers, Mn/DOT directed that four actions be taken with regard to the crack in the girder web:

- Redrill to 6 inches the previously drilled 2-inch-diameter crack-arrest holes.
- Provide a sample of the girder steel to a structural metals engineer for analysis.
- Bolt 3/8-inch-thick steel plates to each face of the web on either side of the diaphragm stiffener.
- Remove the rivets attaching the diaphragms to the stiffeners on each side of the girder and replace them with bolts. Snug, but do not tighten, the bolts (upsetting the bolt threads to keep the nuts from loosening) to add flexibility to the connection.

The Minnesota State bridge engineer recommended that bridge maintenance workers use a core drill to drill 1.5- to 2-inch-diameter holes at each end of the 11 smaller cracks and submit the removed cores for metal analysis. He also recommended that inspectors conduct a "close in-depth" inspection of these areas every 6 months and that they keep a detailed weld/crack inspection log.

By November 1998, Mn/DOT officials had concluded that the cracking in the approach span girders had resulted from stress caused by the rigid connection of the diaphragms to the webs. Because the diaphragms were needed for bracing and could not be removed, an alternative means of attachment was considered. In consultation with the University of Minnesota and after instrumentation measurements and field testing of trial girder retrofits had proven effective, the State bridge engineer, in December 1998, recommended that all the diaphragms near piers 2, 3, 4, 9, and 10 be lowered (brought closer to the bottom flange of

---

[23] None of these cracks were in the deck truss portion of the bridge.

the girder) and attached to the stiffeners with four bolts only. Lowering the diaphragms and changing the attachments were expected to relieve the stress that was causing the cracking. Records indicate that this work was carried out in early 1999.

When the bridge was next inspected, on March 20, 2000, three additional cracks were found in the north approach span girders at the tops of stiffener/diaphragm connections. Workers drilled 1.5-inch-diameter holes at each end of the cracks. As with the previous cracks, these areas were to be inspected at 6-month intervals. In November 2000, based on results of the 6-month inspections of the previous crack repairs, the State bridge engineer recommended that the 12-month inspection cycle be resumed, beginning in 2001. Subsequent fracture-critical inspections revealed additional fatigue cracking in this area, and inspectors recommended that the progression of the cracks be monitored and, when necessary, repaired.

During the 2003 routine bridge inspection and subsequent fracture-critical inspections, cracked welds were noted at numerous locations, primarily on the interior diaphragm tabs. Cracked welds were also noted at various stringer-bearing pedestals. All cracking was in the weld and not in the base metal.

University of Minnesota Fatigue Assessment of I-35W Bridge

In 1999, Mn/DOT commissioned a fatigue study of the I-35W bridge, as it had when fatigue cracking had been found in other bridges in the State. In this case, the agency contracted with the University of Minnesota's Department of Civil Engineering to perform a fatigue assessment of the bridge deck truss.

The assessment involved attaching strain gauges to main truss and floor truss members to measure live-load stress. Researchers monitored the gauges while trucks with known axle weights moved over the bridge. They then developed two- and three-dimensional finite element models of the bridge and used those models to calculate stress ranges throughout the deck truss. No measurements were made on any gusset plates, and the gusset plates were not included in the finite element models.

The March 2001 final report of that assessment concluded that, despite "many poor fatigue details on the main truss and floor truss system":

> The detailed fatigue assessment in this report shows that fatigue cracking of the deck truss is not likely. Therefore, replacement of this bridge, and the associated very high cost, may be deferred.

URS Corporation Inspections and Reports

Fatigue Evaluation. In June 2003, under contract to Mn/DOT, the URS Corporation, of Minneapolis, Minnesota, conducted an initial field inspection of the main truss section of the I-35W bridge. This inspection, which was performed concurrently with Mn/DOT's regularly scheduled bridge inspection, was intended primarily to observe the overall condition of all truss members and to mark bearing and expansion joint positions so that future movement could be measured. URS inspectors also noted the temperature and documented with photographs the general condition of the truss at the time of the inspections.

The "Summary and Recommendations" section of the initial inspection report included the following (abridged):

> The overall condition of the truss members and connections was, from a corrosion standpoint, found to be good. Corrosion was found in localized areas, generally concentrated near the deck joints. Minor corrosion was observed at some of the locations chosen to inspect in the interior of the truss members.
>
> The roller bearings did not appear to be moving freely due to the corrosion, debris and paint build up. The rocker bearings were not accessible for detailed visual observation and assessment of their movement.

In November 2003, URS conducted a second field inspection of the bridge and noted that bearing and joint movement was slight and appeared to be inconsistent between the east and west main trusses. The report indicated that the company would conduct followup monitoring of the bearing and joint conditions and movement at a minimum of five seasonal temperatures. URS returned in January, March, and July 2004 to measure bearing movements.

Fatigue Evaluation and Redundancy Analysis. Despite findings by the University of Minnesota and URS regarding the general condition of the bridge superstructure, Mn/DOT bridge engineers continued to be concerned about the susceptibility of the deck truss portion of the bridge to fatigue cracking, which might lead to failure. Some details that were considered acceptable when the bridge was designed, such as the welds used to connect the diaphragms inside the box member sections, had been subsequently found to be potential sources of fatigue cracking. In December 2003, Mn/DOT contracted with URS to perform a fatigue evaluation and redundancy analysis of the I-35W bridge. The primary objectives of the study were to

> (1) identify critical superstructure members that are most susceptible to cracking, (2) evaluate structural consequences if one of the critical members should sever . . . (3) develop contingency repairs to selected fracture critical members, and (4) establish measures for improving structural redundancy and minimizing tensile stresses in the trusses, and develop a preferred deck replacement staging plan.

URS did not calculate the capacity of the gusset plates based on their actual dimensions "because of the complexities and uncertainties in possible failure sections of the gusset plates and in [the] ultimate capacities of [their] rivets and bolts." Instead, URS assumed that the gusset plates followed the general requirements for connection design contained in the 1961 AASHO (now AASHTO) *Standard Specifications for Highway Bridges* and that they were designed using the AASHTO allowable stress method, which the URS report quoted as follows:

> Connections for main members shall be designed for a capacity based on the average of the calculated design stress in the member and the allowable stress of the member at the point of connection, but in any event, not less than 75% of the capacity of the member.

The URS analysis report stated:

> A review of the [I-35W bridge] plans indicated the likelihood of the use of the 75% member capacity in the truss connection design, since relatively less connection bolts were used for lightly loaded truss members compared with more heavily loaded members of comparable section dimensions.

In its draft report to Mn/DOT in July 2006, URS recommended a retrofit project to strengthen 20 of the 52 fracture-critical main truss members on the bridge. (In January 2007, this recommendation would be revised to include all 52 facture-critical members.) The general objective of the retrofit was to

> replace the strength of a member in an event that the member should completely fail due to a fracture initiated from the concerned fatigue susceptible detail.[24] The added benefit is that the retrofit would also reduce the live load stresses and thus retard or minimize the development of fatigue cracks in the repaired members.

The retrofit identified as "most suitable" was to install steel plates on both side plates of each of the critical box members using high-strength bolts. These plates would "take over all the member forces and replace the lost capacity in the case of a member failure."

In July 2006, according to e-mails and internal documents provided to the Safety Board, Mn/DOT staff began planning to carry out the retrofit project, the cost of which was estimated as $1–1.25 million. In early November 2006, Mn/DOT decided to allocate $1.5 million for the project, with the contract to be let in October 2007 and installation to begin in January 2008 (to allow time to procure the specialized high-tensile-strength steel plates needed for the job).

While planning for the retrofit was underway, some Mn/DOT staff members remained concerned about whether "drilling all those holes in the truss

---

[24] Most of these details were at diaphragm welds inside box member tension members.

box members and terminating the plates at the gusset won't somehow make things worse." One alternative considered was a monitoring system that uses sensors to detect cracks on the critical members. On November 14, 2006, the State bridge engineer indicated that he had discussed this issue with URS, which maintained confidence in the retrofit, and that he still preferred "the certainty of a reinforced member rather than relying on monitoring."

Also during planning for the retrofit, URS was conducting a fracture mechanics analysis of fatigue crack growth on the I-35W bridge. Based on the results of this analysis, URS concluded that the plating retrofit was not necessary and that other "equally viable" alternatives were available for dealing with the fatigue issue.[25] Three alternative approaches were spelled out in a January 2007 executive summary, as follows:

> Steel plating of all 52 fracture critical truss members. This approach will provide member redundancy to each of the identified fracture critical members via additional plates bolted to the existing webs. The critical issue of this approach is to ensure that no new defects are introduced to the existing web plates through the drilled holes. This approach is generally most conservative but its relatively high cost may not be justified by the actual levels of stresses the structure experiences.

> Non-destructive examination (NDE) and removal of all measurable defects at suspected weld details of all 52 fracture critical truss members. The critical issue of this approach is to ensure that no measurable defects are missed by the NDE efforts. The fracture mechanics analysis has indicated that the dimensions of preexisting surface cracks need to be at least one quarter of the web plate thickness in order to grow and subsequently cause member fracture under the traffic load. This approach is most cost efficient.

> A combination of the above two approaches: steel plating of the 24 more fatigue sensitive members . . . and NDE of the 28 more fracture sensitive members.

In January 2007, Mn/DOT decided to perform ultrasonic nondestructive examination of some of the members in the south portion of the deck truss. If inspectors had confidence in the visual inspection and ultrasonic test results , they would continue to the north portion. If not, Mn/DOT would pursue the plating retrofit. In the meantime, Mn/DOT rescheduled the October 2007 retrofit contract for fiscal year 2009, after results of the visual and nondestructive examinations would have been fully evaluated. In March 2007, Mn/DOT entered into a contract with URS to review the results of the inspections and tests and to develop plans and specifications for the plating retrofit if, after the review, that was still considered the best option.

---

[25] None of the members identified in this analysis were believed by the University of Minnesota, by URS, or by Mn/DOT to have stress levels that would likely contribute to fatigue crack growth, but they were members whose failure would have catastrophic consequences.

In May 2007, Mn/DOT inspection teams performed in-depth and nondestructive inspections of half the critical structural members identified by URS. The inspections focused on all 26 members of interest on the west truss and several members on the south end of the east truss. Field notes and photographs from those inspections did not indicate the presence of any significant cracks in those members. A meeting was scheduled between URS and Mn/DOT for August 20, 2007, to discuss the results of the inspections to date and determine whether to continue the inspections or to proceed with retrofit. This meeting was cancelled because of the collapse of the bridge on August 1.

Bowed Gusset Plates on I-35W Bridge

As part of the studies and evaluations performed in 2003 by URS, almost every structural element of the I-35W bridge was documented with photographs. In addition, in 1999, researchers from the University of Minnesota took photographs to document the placement of strain gauges on truss members near the U10 nodes.[26]

Following the bridge collapse, Safety Board investigators reviewed photographs from both of these evaluations that show visible bowing[27] in all four gusset plates at the two U10 nodes. (See figure 19.) At both U10 nodes, the unsupported edges between upper chords U9/U10 and diagonals L9/U10 (south edges of both plates) were bowed toward the west (to the outside of the bridge at U10W and to the inside at U10E). At the two U10' nodes, three of the four plate edges between the upper chords U9'/U10' and the diagonals L9'/U10' (north edges of the plates) were bowed to the east. (The photographs were insufficient to establish the presence of bowing in the remaining gusset plate.) At U10'W, both plates were bowed to the east, toward the inside of the bridge. The west (inside) gusset plate at U10'E was also bowed toward the east. At neither U10 nor U10' did the opposite edges of the plates (north edges at U10 and south edges at U10') appear to be bowed. The U10 and U10' gusset plates were the only plates on the bridge for which the photographs showed obvious evidence of bowing.

---

[26] According to Mn/DOT, the photographs taken by the University of Minnesota were not provided to the agency; and the approximately 225 URS photographs were reduced to 2 x 1.5 inches and reproduced six to a page, on average, in the URS inspection report.

[27] *Bowing* or *bowed*, as used in this report, refers only to the appearance of the gusset plates. These words are not the technical terms that would normally be used by bridge inspectors or the FHWA to describe distortion in a bridge member.

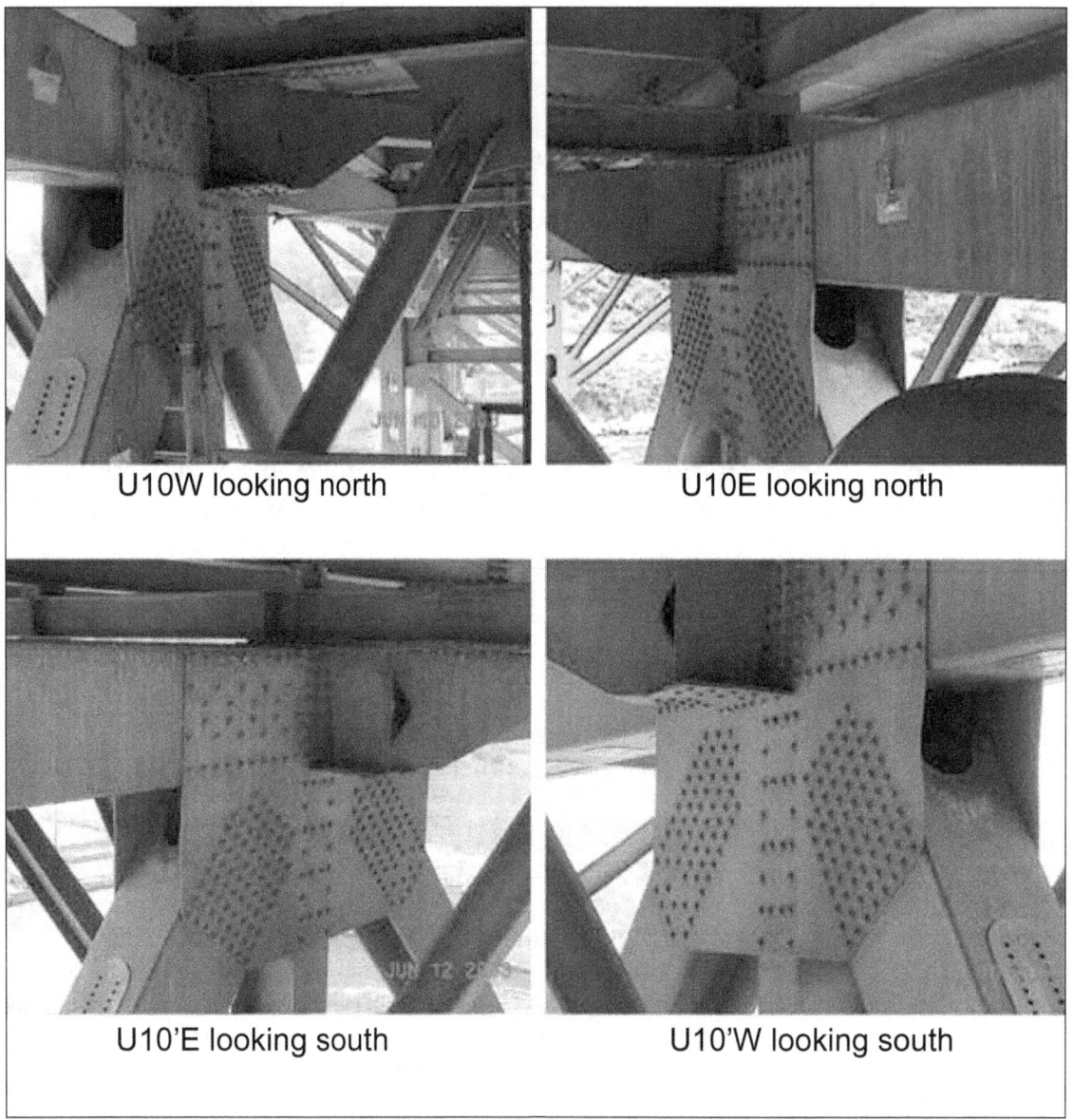

Figure 19. Bowed gusset plates at nodes 10 and 10′. (Source: University of Minnesota, top right photo; URS Corporation, other three photos)

The Safety Board used two methods (referred to as the direct scale method and the dimensional ratio method) to calculate, from the photographs, the maximum displacement of the edges of the four inside gusset plates at each node. The estimated ranges of those calculated displacements are shown in table 6.

Table 6. Calculated displacement of bowed U10 gusset plates.

Inside gusset plate	Displacement (inches)	
	Direct scale method	Dimensional ratio method
U10E	0.72 – 0.81	0.64 – 0.73
U10W	0.47 – 0.64	0.44 – 0.56
U10'E	0.53 – 0.60	0.46 – 0.53
U10'W	0.82 – 0.99	0.80 – 0.94

Neither the University of Minnesota nor URS evaluations made note of the bowed condition of the gusset plates. Review of State bridge inspection records showed that the bowing had not been noted during any previous routine or fracture-critical inspection.

In an interview conducted as part of an independent investigation of the accident by the Minnesota State legislature, the Mn/DOT Metro District bridge safety inspection engineer (a registered professional engineer and a specialist in performing fracture-critical bridge inspections) stated that he had observed the gusset plate bowing during inspections he participated in after joining Mn/DOT in 1997. He said he consulted with another inspector about the bowing and concluded, "that's fit-up, that's construction, that's original construction." He said his first reason for reaching this conclusion was his undergraduate training to the effect that "gusset plates are overdesigned. The safety factors within those gusset plates are 2 to 3."[28] Also, he said, the connection showed no other signs of distress, such as peeling paint, elongation of rivet holes, cracking at welds of the connected members, or cracking or crushing of the bridge deck above the connection. He said the condition was not noted on inspection reports because, "Our inspections are to find deterioration or findings of deterioration on maintenance. We do not note or describe construction or design problems."

The Metro District inspection engineer said that he did not recall when he had first observed the bowing. Comparison of the photographs taken in 1999 and 2003 did not indicate a perceptible change in the magnitude of bowing, though it was not possible for investigators to determine when the bowing first appeared.

Considering the possibility that bowing of the U10 plates may have occurred during erection of the truss (but before addition of the concrete deck), the Safety Board calculated the approximate load in the main truss members connected at the U10 nodes in the two circumstances representing the most severe loading conditions expected during construction (just before and just after closure at the center span), and compared these results with dead loads and design loads for the completed structure as listed in the design drawings. The calculations showed that

---

[28] These safety factors indicate that the gusset plates would have been designed to support 2–3 times the expected loads.

four of the five members in this node had loads in the same direction (compression or tension) but of significantly lower magnitude than either the completed structure dead load or design load. As expected, one member (U10/U12) had a tension load while cantilevered (before closure at the center of the deck truss) but had a compression load after closing of the truss. Both of these loads — tension and compression — were significantly lower in magnitude than the dead load or design load of the completed structure.

Examination of Deck Truss Fracture Areas

As previously noted, at or near the beginning of the collapse sequence, most of the bridge center span fractured and broke away from the rest of the deck truss structure. Video and physical evidence indicated that the breaks in the span occurred just north of pier 6 (south fracture area) and just south of pier 7 (north fracture area) and that the south fracture area break occurred first, followed by the north fracture area break. Safety Board investigators performed a detailed examination of the structural components in the south and north fracture areas. Fractures in these areas were at or adjacent to the U10 or U10' nodes. The salient findings of those examinations are detailed below.

South Fracture Area, Main Trusses

In the south fracture area, the main truss elements that were found to be separated from their nodes, fractured, or damaged included those listed in table 7.

Table 7. Description of damages to members in south fracture area.

Damaged member	Damage description
Diagonals L9/U10	Separated from U10 nodes through gusset plates, bent close to L9 nodes
Chord members U9/U10	Separated from U10 nodes through gusset plates, bent adjacent to U9W and U8E nodes
Chord members L9/L10	Fractured in bending adjacent to L9 nodes, bent close to L10 nodes
Vertical members U10/L10	Compression damage and fractures in upper portion of members above attachment location for lower chord of floor truss 10 and bowing deformation in lower portions of members
Vertical member U9/L9E	Fractured at lower end, separated from U9E node through gusset plates, bent below floor truss
Vertical member U9/L9W	Slight bending damage below floor truss

On both the east and west main trusses, the U9/U10 upper chord member was attached to the south portion of the truss at the U9 end but was fractured from the U10 node through the U10 gusset plates around this member. The diagonal members L9/U10 from both main trusses were attached to the L9 node, and the

U10 end was in the water (west member) or pointing down toward the water (east member). The diagonal L9/U10E did not contain bending deformation directly adjacent to the L9E node but was severely bent starting about 9 feet from the L9E end gusset plate. The diagonal L9/U10W was severely bent adjacent to the L9W node.

The L11 node gusset plates were intact where they connected the lower chord members but had multiple fracture locations in areas above the lower chord.

The lower chord of floor truss 9 remained attached to vertical U9/L9E, and this vertical member had a severe buckle at a location slightly below the lower chord of the floor truss.

U10 Nodes. All four U10 node gusset plates were fractured. As shown in the fracture maps superimposed on CAD (computer-aided design) drawings in figures 20A and 20B, the fracture patterns in the two west plates (inside plate at U10E and outside plate at U10W) were similar, as were the fracture patterns in the two east plates (outside plate at U10E and inside plate at U10W). The gusset plate fractures separated diagonals L9/U10 and upper chords U9/U10 from the remainder of the members in the node. The remaining portion of the gusset plates did not totally fracture; and the upper ends of verticals U10/L10, the U10 ends of diagonals U10/L11, and the upper chords U10/U11 remained connected to each other through at least one of the gusset plates.

Damage patterns on the gusset plates around the U10 ends of diagonals L9/U10 indicated that (1) the gusset plates buckled and bent to the west in the portion of the plate between the L9/U10 diagonal and the upper chord, and fractured mostly under tension loading in the portion of the plate between the L9/U10 diagonal and the U10/L10 vertical; (2) the upper ends of the L9/U10 diagonals shifted laterally to the west relative to the remainder of the nodes; and (3) the remainder of the nodes then moved downward into the L9/U10 diagonals.

The fractures on all four U10 gusset plates left a portion of the plates attached to the upper ends of the separated diagonals L9/U10, with a V-shaped piece of the plates extending beyond the ends of the diagonals. On both the east and west diagonals, these V-shaped pieces were bent toward the east, indicating that the upper ends of the diagonals shifted laterally to the west as the gusset plates deformed and fractured. As these shifts occurred and the remainder of the node dropped, the east side plates of the diagonals penetrated through the interior structure of the nodes. Pieces associated with the remainder of the nodes and floor truss 10 contained multiple impact marks from contact with the upper ends of the diagonals. Deformation patterns associated with a fracture in the upper chord of floor truss 10, just west of the U10E node, were consistent with this fracture having resulted from the impact of the west side plate of diagonal L9/U10E with the lower side of the upper chord of the floor truss.

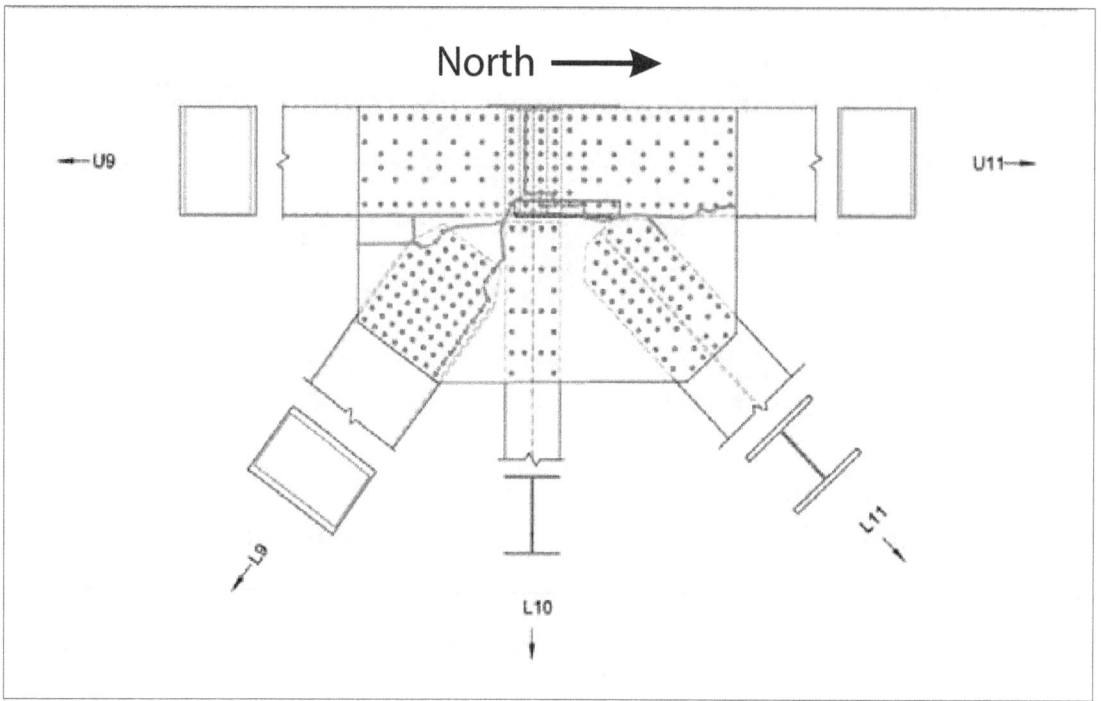

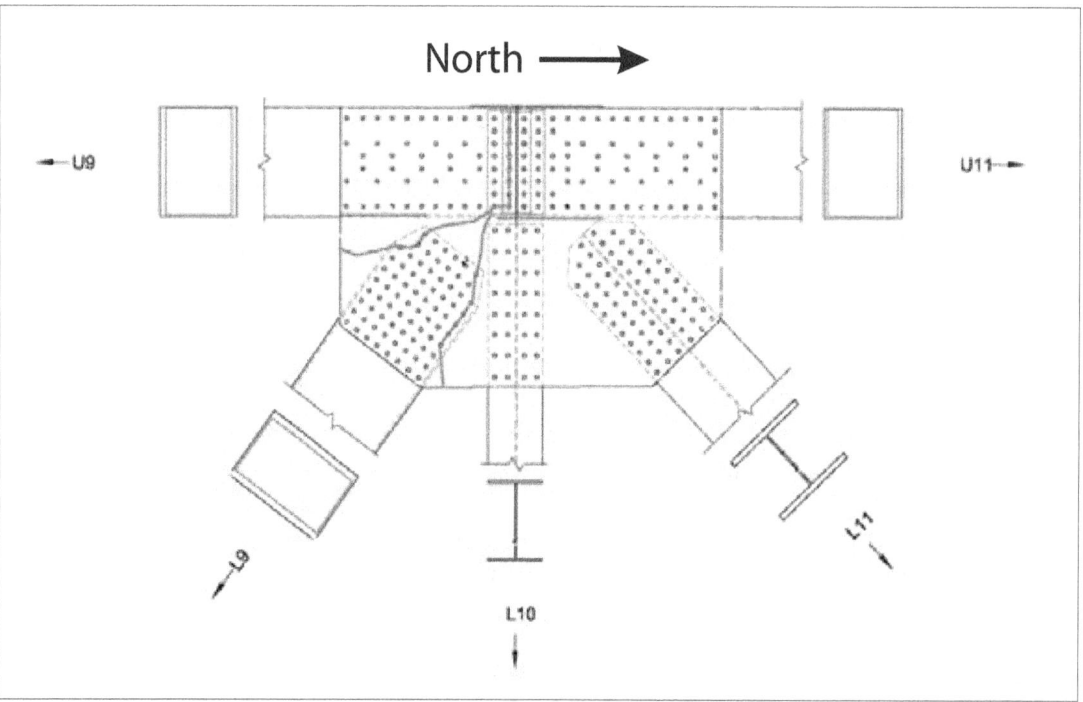

Figure 20A. Inside (east) gusset plate at U10W (top) and outside (east) gusset plate at U10E (bottom), showing similar fracture patterns in blue.

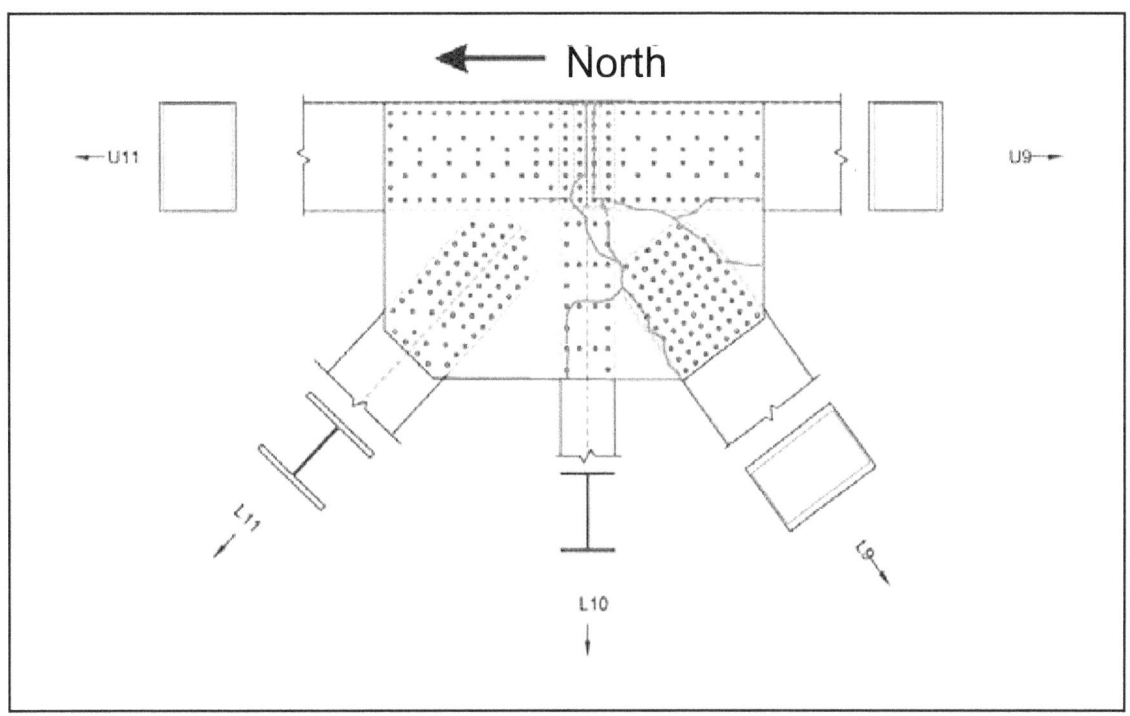

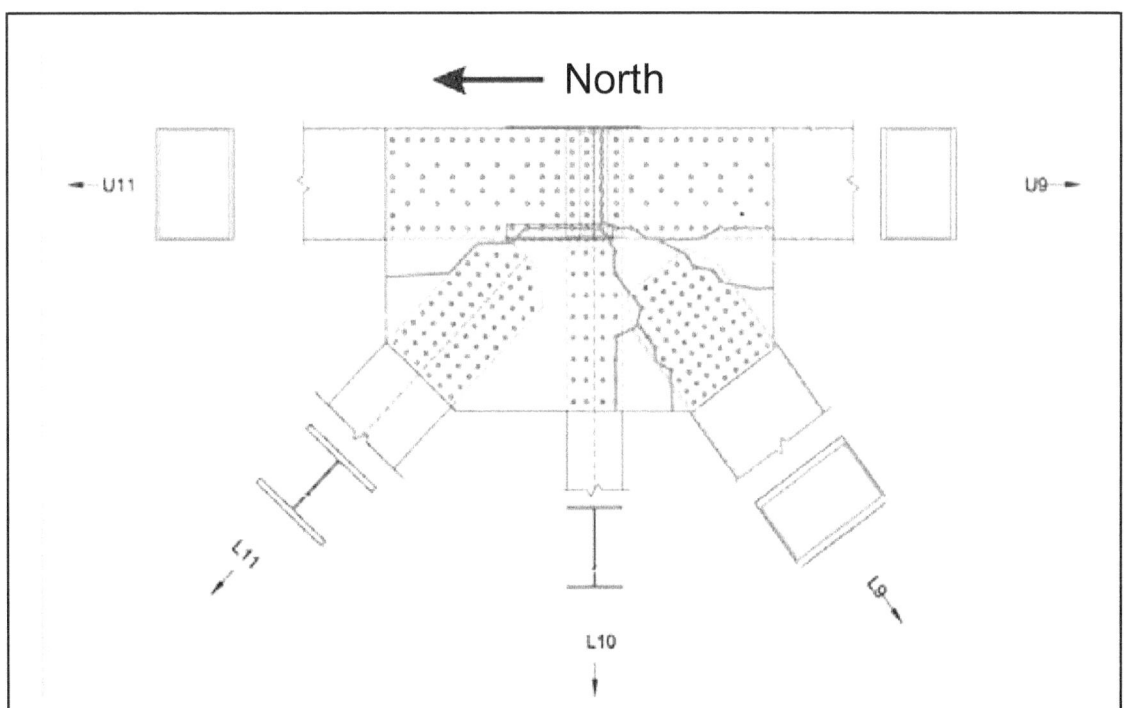

Figure 20B. Outside (west) gusset plate at U10W (top) and inside (west) gusset plate at U10E (bottom), showing similar fracture patterns in blue.

The loss of structural support from the lateral shifting of the L9/U10 diagonals, the bending and fracturing of the U10 gusset plates, and the downward movement of the remainder of the U10 nodes over diagonals L9/U10 caused structural deformations to begin in the deck truss south of these nodes. This deformation would add positive bending loads[29] to the horizontal tension load already present in the remaining portions of the U10 gusset plates across the upper chord. Examination of these fracture areas showed that between each rivet hole, the gusset plate was elongated in a direction slightly offset from horizontal, with the angle gradually increasing from nearly horizontal at the lowest hole to more skewed (down and to the north) at the upper hole. This change in the angle of elongation and the deformation adjacent to the upper edge of the gusset plate were consistent with primarily horizontal tension in the lower portion of the fracture and more shear in the upper portion of the fracture, with the direction of shear indicating that the structure on the north side of the fracture was moving down relative to the structure on the south side of the fracture. The fractures in the portions of the U10 gusset plates on the upper chord were consistent with the expected loading in this area after the U10 nodes began to pull down over diagonals L9/U10. There was no evidence that the U10 gusset plates had any cracking or corrosion before the bridge collapse.

Vertical Members U10/L10. Examination of H-member vertical members U10/L10 revealed significant compression damage, deformation, and fractures in the area above the attachment point for the lower chord of floor truss 10. Except for the brittle fracture just west of the U10E node, the floor truss itself remained largely intact, including the entire area around the U10W node. The webs of both vertical members exhibited impact marks where they were contacted by the east side plates of box-member diagonals L9/U10.

Lower Chord Members L9/L10. Lower chord members L9/L10 from both main trusses were fractured through the northernmost vertical row of rivets at the L9 nodes after significant bending deformation. Examination of these fractures showed evidence of fracture under tension at the top of the members and compression at the bottom of the members, consistent with excessive downward bending. Fracture of the lower chords completely separated the main trusses in the south fracture area. The excessive bending load associated with the fractures in these members indicates that they are secondary events, consistent with downward movement of the L10 nodes after loss of structural support resulting from fractures at the U10 nodes.

Vertical Member U9/L9E. Vertical member U9/L9E was separated from the main truss through fractures of the gusset plates at the U9E node and through fractures of the side plates of the member at the L9E node. The member was also severely bent to the south just below the lower chord of the floor truss. In the postcollapse position, the member was found translated to the west and still

---

[29] Positive bending indicates loading of a horizontal or nearly horizontal member such that the resulting deformation deflects the member toward the shape of the letter "U."

attached to the lower chord of floor truss 9. The damage patterns associated with the fractures at the upper and lower ends of the member indicated that the center portion of the member moved to the south relative to the upper and lower ends, consistent with the southward bend of the vertical member slightly below floor truss 9. The lower chord of floor truss 9 in the segment adjacent to main truss vertical member U9/L9E was twisted in a direction that would impart a bending moment into the vertical member, which is also consistent with the southward direction of the bend below the floor truss.

The fractures and deformations in vertical U9/L9E were consistent with the member initially bending to the south at the location below the floor truss, followed by fracturing at the upper and lower ends of the member as the bending deformation increased. Finite element analysis (discussed in the "Tests and Research" section of this report) showed that, at the time of the bridge collapse, vertical members U9/L9 in both main trusses were loaded in compression below the floor truss, but the loads were not sufficient to cause buckling.

Vertical Member U9/L9W. Vertical member U9/L9W remained attached to the west main truss at both its upper and lower ends. This member contained a slight bend to the south just below the lower chord of floor truss 9 (the same location as the severe bend in vertical member U9/L9E). The lower chord of floor truss 9 in the segment adjacent to U9/L9W was twisted in a direction that would impart a bending moment into the vertical member, which is also consistent with the southward direction of the bend below the floor truss. The amount of twisting in this segment of the floor truss lower chord was less than the amount of twisting in the corresponding segment of the lower chord adjacent to U9/L9E.

Upper Chord Members U9/U10W and U8/U10E. With continued dropping of the center portion of the deck truss, the upper chords south of the U10' nodes were bent down, with eventual negative bending deformation in the upper chord members adjacent to the U8E node (for upper chord member U8/U10E) or the U9W node (for upper chord member U9/U10W). The lack of significant bending deformation in the upper chord adjacent to the U9E node indicates that vertical U9/L9E was separated from the east truss or bent before the upper chord member from U8 to U10 moved downward.

Diagonals L9/U10. Also, with continued dropping of the center portion of the deck truss, diagonals L9/U10 on both main trusses—which were fractured from the U10 nodes but still connected at the L9 nodes—were displaced laterally and bent down, with eventual negative bending deformation in these members adjacent to or near the L9 nodes.

South Fracture Area, Deck and Stringers

Examination of the postcollapse position of the bridge structure showed that the expansion joint in the deck at floor truss 14 did not appreciably open. At floor truss 12, the deck stringers were bent downward, and the deck was fractured at what appeared to be a construction joint. The deck south of this construction joint had displaced to the south relative to the underlying stringers. Floor truss 11 and the stringers and deck above it were underwater. Three layers of deck and stringers—from nodes 11 to 10, from nodes 10 to 9, and from nodes 9 to the expansion joint at nodes 8—were stacked in a folded pattern in what would have been the truss panel between nodes 10 and 11. These layers were more irregularly folded on the southbound deck slab compared to the northbound slab. Based on the visible portion of the deck, the deck and stringers had a positive bend approximately above floor truss 10 and a negative bend above floor truss 9. The deck had an expansion joint at nodes 8, and the deck and stringers above the south fracture area were only bearing on floor truss 8 without any mechanical fasteners.

The location of the positive bend in the deck above floor truss 10 is consistent with the direction of loading placed on the stringers as the deck and stringers resisted the downward movement of the floor truss following separation of the L9/U10 diagonals from the U10 nodes. Simultaneously, as the deck and stringers at floor truss 10 were pulled down, the deck and stringers above floor truss 9 were loaded in negative bending. The postcollapse location of the deck and stringers above the south fracture area indicates that the deck and stringers from nodes 12 south to nodes 8 separated from the main trusses, and that the deck and stringers from nodes 10 to nodes 8 transitioned to the north as the collapse progressed, resulting in the folded pattern in the postcollapse position.

South Fracture Area, Floor Trusses and Braces

Floor truss 8 remained at least partially attached to the vertical members in the east and west main trusses but was significantly damaged during later stages of the collapse by impact with the ground and pier 6.

The portion of floor truss 9 between the main trusses (the central portion) was attached to vertical U9/L9E through the lower chord of the floor truss, and this vertical member had a severe bend to the south at a location slightly below the lower chord of the floor truss. As previously discussed, vertical U9/L9E was fractured from both the upper and lower nodes and displaced toward the centerline of the pier. From its attachment to vertical U9/L9E, the central portion of floor truss 9 extended longitudinally north, bending down into the river. The central portion of floor truss 9 was separated from vertical U9/L9W through the gusset plate at the lower floor truss chord attachment and through a fracture in the floor truss upper chord at the upper chord of the west main truss.

The upper chord of floor truss 9 was deformed to the north in two lobes on each side of center. Near its middle, the central portion of this upper chord had a corresponding slight bend to the south, consistent with this portion of floor truss 9 being restrained by the upper lateral brace members that attach to the south side of the center of the upper chord, as the stringers pulled adjacent portions of the upper chord to the north.

The central portion of floor truss 10 remained partially attached to portions of main truss verticals U10/L10 through the lower chord of the floor truss. The vertical members were found lying on the lock guide wall with their upper ends in the water and the central portion of floor truss 10 between the verticals approximately directly below its position in the bridge, with the west vertical farther south than the east vertical. Postcollapse, the central portion of floor truss 10 was partially submerged.

The upper chord of floor truss 10 had a deformation pattern similar to but more pronounced than the pattern found on the upper chord of floor truss 9. The pattern was consistent with the deck stringers pulling the upper chord northward, with the upper chord restrained at the main trusses and by the upper lateral braces attaching to the south side of the upper chord (in the center). The upper chord of floor truss 10 was also fractured near the U10E node, and part of this fracture was brittle.

Floor truss 11 was heavily damaged and was recovered approximately directly below its normal position. It was partially covered by the three folded layers of deck and stringers. Damage to floor truss 11 was consistent with crushing damage during the final stages of river impact. None of the floor trusses in the south fracture area exhibited a fracture indicative of a pure tension or pure compression failure.[30]

Most of the lateral and sway brace members between nodes 8 and 11 were present in the jumble of members on the south riverbank. Deformations associated with the upper lateral braces between floor trusses 8 and 9 and between floor trusses 9 and 10 were consistent with damage occurring as a result of resisting the northward movement of floor trusses 9 and 10.

North Fracture Area, Main Trusses

The deck truss remained relatively intact south of the U10' nodes (in the center portion that fell into the river) and in the rigid body portion that rotated north on pier 7. Thus, the north fracture area was north of nodes 10' and south of the U8' and L9' nodes. In the north fracture area, certain main truss elements were found to be separated from their nodes, fractured, or damaged. (See table 8, which also includes damage to lower chord members L10'/L11', which were outside the fracture area as defined here.)

---

[30] A member that fractured under pure tension loads would not have bending deformation, and a member that failed under pure compression loads would be expected to buckle approximately at the midpoint of its length.

Table 8. Description of damages to members in north fracture area.

Damaged member	Damage description
Diagonals L9'/U10'	Separated from U10' nodes east and west through gusset plates, bent or fractured close to L9' nodes east and west
Upper chord members U9'/U10'	Separated from U10' nodes east and west through gusset plates, bent adjacent to U8' nodes east and west
Vertical members U10'/L10'	Compression damage and fractures in upper portion of members above attachment location for lower chord of floor truss 10', and bowing deformation in lower portions of members
Lower chord members L9'/L10'	Fractured in bending adjacent to L9' nodes east and west
Lower chord members L10'/L11'	Compression buckling adjacent to L11' nodes
Vertical members U9'/L9'	Fractured at lower end, separated or nearly separated from U9' nodes through gusset plates, bent below floor truss

As previously discussed, the video recording shows that the fracture that occurred near the north end of the bridge center span was secondary to the fracture occurring near the south end.

The locations and types of fractures and damage were generally similar for both the east and west main trusses, with two exceptions. The L9' end of diagonal L9'/U10'W fractured from severe bending adjacent to the node, allowing this diagonal to fall into the river, while the L9'E end of diagonal L9'/U10'E was severely bent but did not fracture and remained attached to the L9'E node. Also, vertical U9'/L9'W completely separated from the main truss at both ends, while the corresponding vertical on the east truss remained minimally attached at the L9'E node.

U10' Nodes. All four gusset plates at the U10' nodes were fractured and deformed in a manner similar to the gusset plates at the U10 nodes, with the fractures separating the diagonals L9'/U10' and upper chords U9'/U10' from the nodes. On both the east and west main trusses, the other U10' members (upper ends of verticals U10'/L10', U10' ends of diagonals U10'/L11', and upper chords U10'/U11') remained connected to each other through at least one of the gusset plates.

Damage patterns on the gusset plates around the U10' ends of diagonals L9'/U10' indicated that (1) the gusset plates buckled and bent in the portion of the plate between the L9'/U10' diagonal and the upper chord, and fractured mostly under tension loading in the portion of the gusset plate between the L9'/U10' diagonal and the U10'/L10' vertical; (2) the upper ends of the L9'/ U10' diagonals shifted laterally to the inside of the bridge (east for diagonal L9'/U10'W and west for diagonal L9'/U10'E) relative to the remainder of the nodes; and (3) the remainder of the nodes then moved downward into the L9'/ U10' diagonals.

The deformation and fracture patterns in the U10′ node gusset plates associated with the upper ends of the L9′/U10′ diagonals indicate that these upper ends shifted laterally to the inside of the bridge relative to the remainder of the nodes. Pieces associated with the remainder of the nodes and floor truss 10′ contained multiple marks from impact with the upper ends of the diagonals.

The fractures through the portions of the U10′ gusset plates that remained with the separated upper chords (U9′/U10′) were similar to corresponding fractures in the U10 gusset plates (figures 20A and 20B), with separation primarily through the first row of rivets north of the node centerline. Also, like the U10 node gusset plates, the U10′ gusset plates in the area of the upper chord fractured primarily under horizontal tension loads in the lower portion of the fracture and under more shear loads in the upper portion of the fracture, with the direction of shear indicating that the structure on the south side of the fracture was moving down relative to the structure on the north side of the fracture. The remaining portions of the gusset plates on the U10′ nodes remained intact and contained much less distortion than corresponding portions of the gusset plates on the U10 nodes.

The upper chords north of the U10′ nodes were bent downward, with eventual negative bending deformation in the upper chord members adjacent to the U8′ nodes. The lack of significant bending deformation in the upper chord adjacent to the U9′ nodes indicates that verticals U9′/L9′ were separated from the trusses or buckled before the upper chord members from U8′ to U10′ were pushed downward.

Vertical Members U10′/L10′. Verticals U10′/L10′ were both bent or buckled in several locations: directly below the U10′ nodes, below the lower chord of floor truss 10, just below the midstrut attachment location, and above the L10′ node gusset plates. Although vertical U10′/L10′W contained significant damage in the area below the U10′W node, this area remained at least partially intact throughout the collapse, unlike corresponding portions of the U10/L10 verticals in the south fracture area.

Lower Chord Members L9′/L10′. Lower chord members L9′/L10′ were both fractured in negative bending adjacent to the L9′ nodes, consistent with downward motion of the center portion of the deck truss. These fractures were similar to fractures in corresponding members in the south fracture area (L9/L10). Lower chord L9′/L10′W also had a compression buckle 15 feet north of the L10′W node.

Lower Chord Members L10′/L11′. Lower chord members L10′/L11′ had large compression buckling areas adjacent to the L11′ nodes. In these areas, the upper and lower cover plates were partially fractured from the side plates, and the side plates formed a large "S" shape, with deformation primarily in the horizontal plane. Lower chord member L10′/L11′ also had compression buckling adjacent to the L10′E node. The compression damage to lower chord members L10′/L11′ east and west and L9′/L10′W was consistent with high compression loads generated in the lower chord as the center portion of the deck truss dropped toward the river in the south fracture area.

Vertical Members U9'/L9'. The two vertical members U9'/L9' had similar damage and fractures except that the lower end of U9'/L9'E remained partially attached to the L9'E node, while the corresponding area on U9'/L9'W was completely fractured from the L9'W node. Both of these members contained bending damage just below the lower chord of floor truss 9' (with the area of bending damage displaced to the north) and severe bending damage at the ends. The damage on these two members was similar to the damage on vertical U9/L9E from the south fracture area.

North Fracture Area, Deck and Stringers

The deck and stringers above the north fracture area collapsed without the folding that was noted in the deck and stringers above the south fracture area. The deck and stringers came to rest directly below their positions in the bridge, indicating that they became separated from floor truss 8' and that the structure between nodes 8' and 10' fractured and separated sufficiently to allow the deck to drop almost vertically.

Floor Trusses and Braces. Most of floor truss 10' was found in the river but substantially intact. Between verticals U10'/L10', the truss was bowed northward, consistent with restraint being provided by the upper lateral system between the U9' nodes and the upper lateral attachment point on the north side of the center of floor truss 10'.

The top chord of floor truss 9' between the primary truss verticals to which it was attached was bowed southward. The direction of this bowing was consistent with the deck stringers pulling on the chord southward toward the river after the upper lateral brace became separated from the north side of the center of the floor truss. None of the floor trusses in the north fracture area exhibited a fracture indicative of a pure tension or pure compression failure.

Tests and Research

Preexisting Features of Structural Components

While examining and documenting the condition of structural components of the I-35W bridge, Safety Board investigators also noted damage or conditions, such as corrosion or cracking, that predated the collapse. Corrosion with accompanying section loss was noted in all four gusset plates at the L11 nodes (a site where corrosion and section loss had been noted on a 1993 Mn/DOT bridge inspection report). The line of corrosion extended across the inside faces of the gusset plates, approximately along the edges of top cover plates of the box member chords to which the gusset plates were attached. Where it was possible to do so, investigators measured the amount of section loss at the top of the lower chord of the main truss along the 99-inch width of the gusset plate. Field measurements were taken at 1-inch intervals using an electronic point micrometer and were later confirmed using laser scanning of the corroded areas.

Thickness data from some portions of the L11 gusset plates could not be obtained either because the area of interest was not recovered following the bridge collapse or was inaccessible because of damage or deformation. The least amount of corrosion was on the L11W node outside (west) gusset plate, and the most was on the L11E node inside (west) and L11W node inside (east) gusset plates. The west gusset plate at the L11E node was cited in the 1993 inspection report as having "loss of section 18" long and up to 3/16 [0.1875 inch] deep." As documented after the collapse, this gusset plate had a line of corrosion along almost its entire length, with an average loss of 17 percent. Table 9 shows the results of the laser scan measurements of the L11 gusset plates.

Table 9. Statistics for laser scan measurements of L11 gusset plate thickness.

Measurement	L11E node outside (east) gusset plate	L11E node inside (west) gusset plate	L11W node inside (east) gusset plate	L11W node outside (west) gusset plate
Mean thickness (inches)	0.477	0.414	0.421	0.452
Minimum thickness (inches)	0.362	0.274	0.276	0.340
Mean section loss (%)	4.7	17.1	15.8	9.6

Light-to-moderate rust and corrosion were found on various other structural members, as had been documented in Mn/DOT inspection reports. Areas of preexisting fatigue cracking in the multigirder approach spans showed no change in crack length since the 2006 Mn/DOT fracture-critical bridge inspection.

I-35W Bridge Gusset Plate Adequacy Analysis

The FHWA participated in the investigation of this accident. As part of the investigative effort, FHWA engineers reviewed and assessed the design of the main truss gusset plates used on the I-35W bridge. The findings of that review and assessment are contained in a report, *Adequacy of the U10 Gusset Plate Designs for the Minnesota Bridge No. 9340 (I-35W Over the Mississippi River)*, and are summarized below.

The general notes on the construction drawings for the I-35W bridge indicate that Mn/DOT commissioned the design to meet division I of the AASHO *Standard Specifications for Highway Bridges*, 1961 edition, and the 1961 and 1962 *Interim Specifications* as modified by Minnesota Highway Department standards n allowable stresses. The AASHO specifications included the statement that "gusset plates shall be of ample thickness to resist shear, direct stress, and flexure, acting on the weakest or critical section of maximum stress."

Neither Mn/DOT nor Jacobs Engineering was able to locate the original calculations that were done in design of the I-35W bridge main truss gusset plates. Using a basic design methodology consistent with that used by Sverdrup & Parcel to design the gusset plates for the floor trusses,[31] FHWA engineers calculated the stresses on the main truss gusset plates that would be generated by the design loads (demand) in the members secured by the gusset plates. Comparing these stresses to the allowable stresses (capacity) in the AASHO specifications resulted in demand-to-capacity (D/C) ratios that illustrate the expected performance of the gusset plates. The D/C ratio is a comparative measure of the efficiency of the design. A D/C value of less than 1 (meaning that maximum potential design stress, or demand, is less than design capacity[32]) indicates a conservative design; a D/C ratio of 1 (demand is equal to design capacity) indicates an efficient design; and a D/C ratio of greater than 1 (demand exceeds design capacity) indicates a liberal design. Liberal designs are not common but are sometimes acceptable based on the professional judgment of an engineer. D/C ratios that are significantly greater than 1 can also indicate a design error.

D/C ratios were calculated for all the I-35W bridge main truss gusset plates except for those at the U0 and L8 nodes, which were of a significantly different configuration than the other main truss nodes. The evaluations considered two critical sections in each gusset plate—one horizontal section near the center of the gusset along the edge of the chord member, and one vertical section adjacent to the vertical member of the node. Shear, principal tension, and principal compression calculations were performed along each section. These calculations showed that the gusset plates at the U4('), U10('), and L11(') nodes had D/C ratios for shear that exceeded 2, and had D/C ratios for principal tension and principal compression that exceeded 1, sometimes by a substantial amount. In addition, the gusset plates at two other nodes had D/C ratios slightly over 1 for shear. The gusset plates at the U4('), U10('), and L11(') nodes provided about one-half of the resistance required by the design loadings.

The AASHO specifications also required that if the length of an unsupported edge[33] of a low-alloy steel gusset plate exceeds 48 times its thickness, the edge shall be stiffened. All 24 gusset plates at the U4('), U10('), and L11(') nodes east and west were 0.5 inch thick. The length of unsupported and unstiffened edge at U10(') measured 30 inches, which exceeded the allowable maximum of 24 inches (48 x 0.5 inch). Safety Board investigators examined the original bridge design drawings and specifications and determined that the 0.5-inch-thick gusset plates at the U4('), U10('), and L11(') nodes were fabricated and installed in accordance with the original plans.

---

[31] Calculations for the welded floor truss gusset plates were found in the packages of computation sheets retained by both Mn/DOT and Jacobs.

[32] Design capacity incorporates a margin of safety by using an allowable stress that is significantly less than the minimum specified yield stress of the member material, such that the ultimate capacity of a component is expected to be significantly greater than its design capacity.

[33] The *unsupported edge* of a gusset plate is that portion of the plate that is not directly attached to a structural member.

On January 11, 2008, the FHWA provided an interim report on the adequacy of the gusset plates. Based on the findings in that interim report and the Safety Board's examination of the failed structure—and in the interest of possibly preventing a similar catastrophic failure even while the investigation of this accident was underway—the Safety Board, on January 15, 2008, issued the following safety recommendation to the FHWA:

H-08-1

For all non-load-path-redundant steel truss bridges within the National Bridge Inventory, require that bridge owners conduct load capacity calculations to verify that the stress levels in all structural elements, including gusset plates, remain within applicable requirements whenever planned modifications or operational changes may significantly increase stresses.

In coordination with the issuance of Safety Recommendation H-08-1, the FHWA issued Technical Advisory T 5140.29, *Load-carrying Capacity Considerations of Gusset Plates in Non-load-path-redundant Steel Truss Bridges*, on January 15, 2008. The technical advisory referenced Safety Recommendation H-08-1 and recommended that bridge owners take the following actions to supplement the guidance in the AASHTO *Manual for Condition Evaluation of Bridges*:

- **New or replaced non-load-path-redundant steel truss bridges.** Bridge owners are strongly encouraged to check the capacity of gusset plates as part of the initial load ratings.

- **Future recalculations of load capacity on existing non-load-path-redundant steel truss bridges.** Bridge owners are strongly encouraged to check the capacity of gusset plates as part of the load rating calculations conducted to reflect changes in condition or dead load, to make permit or posting decisions, or to account for structural modifications or other alterations that result in significant changes in stress levels.

- **Previous load ratings for non-load-path-redundant steel truss bridges.** Bridge owners are recommended to review past load rating calculations of bridges which have been subjected to significant changes in stress levels, either temporary or permanent, to ensure that the capacities of gusset plates were adequately considered.

In an April 30, 2008, letter to the Safety Board in response to Safety Recommendation H-08-1, the FHWA referenced the technical advisory and indicated that, after its promulgation, the FHWA and AASHTO had worked together to provide technical assistance and guidance to FHWA field offices, bridge owners, and State DOTs in evaluating load ratings and gusset plates on steel truss bridges. Based on this initial response, the Safety Board, on July 23, 2008, classified Safety Recommendation H-08-1 "Open—Acceptable Response."

Analysis of Other Aspects of I-35W Bridge Design

The "general notes" in the design drawings for the I-35W bridge indicated that the design was to be in accordance with the Minnesota Department of Highways *Standard Specifications for Highway Construction*, dated January 1, 1964, along with division I of the AASHO *Standard Specifications for Highway Bridges*, 1961 edition, with 1961 and 1962 *Interim Specifications*. These specifications dictated the allowable stresses used for the design.

Following the bridge collapse, the Safety Board received design documents from both Mn/DOT and Jacobs Engineering. In addition to the design drawings, they provided sequentially numbered and indexed compilations of checked computation sheets showing the calculations performed for the final design. These computation sheets covered the superstructure (226 pages) and substructure (56 pages) for the 11 approach spans and the superstructure (223 pages) and substructure (136 pages) for the deck truss portion of the bridge. Jacobs Engineering also provided records showing work on preliminary designs.

The computation sheets for the deck truss superstructure contained calculations for the welded floor truss gusset plates, including a set of calculations showing three iterations to arrive at the final design of the U3 floor truss node. None of the records that had been retained by Mn/DOT or Jacobs Engineering showed any calculations for the nodes on the main trusses, which would have been needed to determine the required numbers of rivets and rivet patterns as well as the size and thickness of the gusset plates and splice plates at each node. Jacobs Engineering did locate a former Sverdrup & Parcel employee who had personally retained some unchecked computation sheets for the preliminary design of the bridge. As discussed later in this report, these sheets showed the calculations used to determine the rivet patterns and the required numbers of rivets for the gussets and to size the plates, but the calculations considered only the chord carry-through forces and not the forces from the diagonals. The U10 and L11 gusset plates in those preliminary calculations were the same thickness and material as in the final design used on the bridge.

As part of its gusset plate adequacy analysis, the FHWA reviewed the methodology that Sverdrup & Parcel had used to design the deck truss portion of the bridge, as documented in the computation sheets. The method used to analyze the statically indeterminate main truss was checked and found to be correct. The assumptions made regarding the magnitude, distribution, and transfer of the dead loads were appropriate, and those calculations were correct. The live load calculations appropriately followed the AASHO specifications for lane loading and impact, and the live loads in the members connecting at the U10 nodes were checked and found to be correct. The allowable stresses used for design of the members (tension and compression) followed the AASHO specifications. The calculations for the welded floor truss gusset plates were judged to be consistent with examples of gusset plate analyses from other sources.

The FHWA also enlisted a bridge design contractor, Bridge Software Development International, to evaluate the original design of the deck truss portion of the I-35W bridge using proprietary finite element analysis software that has the capability to incorporate the AASHO-specified live loads in the analysis. This finite element model used beam elements and truss elements (truss elements have no bending degrees of freedom) to model the main trusses and floor trusses, with the beam elements connecting at a point (the model did not include any details of the connections, such as gusset plates). Beam and shell elements were used to model the stringers, with solid elements used for the deck. The bearings at piers 5, 6, and 8 were modeled as moving freely north and south.

The evaluation report concluded that the original design accurately predicted dead load and live load forces in the main trusses. The results for the floor trusses were dependent on the details of the model: If the deck structure was assumed to act in concert with the floor truss, the loads in the original design plans were very conservative; if the deck structure was removed, the combined dead loads, live loads, and impact loads in some members in the upper chord of the floor truss were somewhat higher than the loads shown in the original design plans.

In summary, with the exception of the main truss connections (including the gusset plates), no significant deficiencies were identified in the overall methodology or calculations performed to design the deck truss portion of the bridge. The documentation for other aspects of the bridge design was sufficient to reconstruct the original assumptions and check the calculations, and the methodology was consistent with the AASHO specifications.

Materials Testing

To test the material properties of the four gusset plates from the U10E and U10W nodes, a rectangular section was cut from an undamaged area of each gusset plate still riveted to the U9/U10 upper chord members. The FHWA performed tensile tests, Charpy V-notch tests,[34] and compact tension fracture toughness tests on samples from all four sections, with samples cut so as to be aligned either north–south or up–down to assess variations in material properties that might have been introduced during manufacture. The measured yield stresses of all of the tensile test samples exceeded the final design plan specified minimum yield stress of 50,000 psi. The test samples for the Charpy V-notch and fracture toughness tests all exhibited a ductile (as opposed to brittle) mode of fracture, and the Charpy V-notch tests satisfied current AASHTO requirements (developed after the I-35W bridge was built) for bridge steels. For all four gusset plates, the tensile tests showed no sensitivity to the sample orientation, but the Charpy V-notch tests and the fracture

---

[34] In *Charpy V-notch* testing, a falling pendulum strikes a rectangular specimen that has a V-shaped notch in the middle and is supported at each end. The test measures the amount of energy (typically in ft-lbs) required to fracture a specimen.

toughness tests indicated that samples with cracks oriented up–down had higher toughness than samples with cracks oriented north–south, indicating a generally north–south rolling direction during manufacture of the gusset plates.

The FHWA also performed tensile tests on samples taken from main truss members and from floor truss FT10. The main truss members sampled included both upper chord members and both diagonals at all four of the U10 and U10' nodes (east and west), as well as the four L9/L11 lower chord members directly below those nodes. For the main truss members, four samples were tested from the box members (one each from the two side plates and two cover plates), and three samples were tested from the H members (one each from the two side plates and one from the web plate). For FT10, samples were taken from five of the different-sized rolled wide flange sections that made up the floor truss, with a sample taken from the two flanges and the web at each location. All of the tensile test samples were oriented along the axis of the member. The measured yield stresses of most of the samples exceeded the specified minimum yield stress, but some did not. The FHWA report concluded that this variation was not abnormal and that it would be accounted for by the conservative philosophy of bridge design specifications.

The Safety Board performed hardness tests on samples from the four gusset plates from the U10E and U10W nodes and from the four gusset plates from the L11E and L11W nodes. The hardness of the L11 gusset plates exceeded the hardness of the U10 gusset plates by an average of 6 percent. Because yield stress and ultimate strength correlate with hardness for steels, these results—coupled with the tensile tests of the U10 gusset plates—indicate that the L11 gusset plates also had yield stresses above the specified minimum.

Cores were taken from the piers and from the bridge deck to measure the density, compressive strength, stiffness, and coefficient of thermal expansion of the concrete. The lengths of the cores from the bridge deck were also measured to determine the distribution of concrete thickness along the bridge. Tests were performed by the FHWA and by Mn/DOT and its consultant, Wiss, Janney, Elstner Associates, Inc. All of the compressive strengths measured exceeded those required, and the results of the other tests were considered to be within normal ranges.

Finite Element Modeling of I-35W Bridge

The Safety Board, in September 2007, began working in collaboration with the FHWA, the State University of New York (SUNY) at Stony Brook, and Dassault Systemes Simulia Corporation (Simulia) to develop finite element models[35] to evaluate the forces and deformations in the I-35W bridge at the time of the collapse. Information gained from examining the wreckage was used to guide the

---

[35] A *finite element model* is a computer model describing a virtual assembly of simplified structural elements used to approximate a complex structure. The behavior of the complex structure is then calculated by combining the actions of the interconnected simpler elements.

modeling effort and evaluate results. Based on findings from the accident scene, the modeling was initially focused on the U10 nodes. The modeling effort was expanded to include the L11 nodes as a result of the FHWA's gusset plate adequacy analysis, which showed that the gusset plates at both the U10 and L11 nodes had inadequate load capacity; and the fact that the L11 gusset plates were found to have areas of corrosion, which further reduced their load-carrying capacity.

The models were based on the original Sverdrup & Parcel design plans and the Allied Structural Steel shop drawings. The models also reflected the various modifications the bridge had undergone. Data obtained from analysis of the wreckage materials were also incorporated into the models, including the physical properties and dimensions of the concrete deck and the steel truss and gusset plates. All model calculations were performed using Abaqus software.

The FHWA developed a three-dimensional global model of the entire deck truss portion of the bridge, constructed with beam and shell elements. The global model, including the boundary conditions at the piers, was calibrated using the live load strain gauge data obtained by the University of Minnesota as part of its 1999 fatigue assessment of the I-35W bridge.

In addition, both the FHWA and SUNY/Simulia created detailed models of the U10 and L11 nodes to examine the fine details of stress distribution. These detailed models were built into the FHWA global model of the bridge. The FHWA detailed model generally used shell-element representations for the truss members and the gusset plates, while the SUNY/Simulia detailed models used solid-element representations. The two approaches generally gave similar results, though some details differed.

The models of the bridge were altered in steps to represent the structure at various stages of modification since its opening in 1967. These steps included the as-built bridge, the increase in weight associated with the 1977 increase in deck thickness, the increase in weight associated with the 1998 change in the outside traffic railings and median barrier, and the decrease in weight associated with the removal of part of the deck in the repaving operation that was underway at the time of the collapse. To explore the conditions of the bridge on August 1, 2007, the additional effects of the weight of traffic, the weight of construction loads, and the temperature changes that day were also incorporated. The models included the bowing deformation of the U10 node gusset plates that was evident in photographs taken in 1999 and 2003. The estimate of the bowing magnitude used for input to the models was 0.60 ± 0.15 inch. To generate bowing consistent with that shown in the photographs, it was necessary to use an initial maximum deflection of 0.5 inch in the unloaded model, which increased to 0.6–0.7 inch after application of he original bridge dead load plus the added deck and the modified barriers (the loading condition at the time the bowed gusset plates were photographed).

In the first step, under only the dead load of the original bridge design, the models showed localized areas of yielding[36] in both the U10 and L11 gusset plates. The area of yielding expanded as weight was added to represent the 1977 and 1998 modifications. This finding is consistent with the FHWA's identification of the inadequacy of the U10 and L11 gusset plates with respect to their design requirements. The general intent of those design requirements should have kept the stress in the gusset plates below 55 percent of the yield stress under the design loads for the members connected. The design load includes the dead load plus an AASHO-specified live load. The models also showed that the stresses in all of the upper and lower chord members and in all of the vertical and diagonal members of the bridge were within acceptable ranges under all conditions evaluated.[37]

The yielding in gusset plates from the U10 nodes was generally found at the ends of the diagonal members. At node U10W, the maximum stress in the gusset plates occurred between the upper end of compression diagonal L9/U10W and upper chord member U9/U10W. Figure 21 presents stress contour plots generated from the finite element analysis that show how the area of yielding expanded from the as-built condition to the day of the bridge collapse.

The detailed models of U10W showed that, at lower loads, the areas of yielding were surrounded and constrained by material below the yield stress, preventing appreciable displacements of the gusset plates or truss members. As the load increased due to successive bridge modifications and added live loads, the yielding in the gusset plates spread through a wider area, and the upper end of diagonal L9/U10W began to shift laterally. In the condition just before collapse, almost all of the gusset plate material surrounding the upper end of diagonal L9/U10W was yielding, and the lateral shift of the diagonal became more pronounced. (See figure 22.)

Under further increasing loads, the lateral displacement of the upper end of diagonal L9/U10W eventually became unstable.[38] The calculated loads required to trigger this lateral shifting instability in the models were slightly higher than the estimates of the bridge dead load plus traffic and construction live load. These differences are considered to be within the range of approximations inherent in the model and the uncertainties associated with the magnitude and distribution of the bridge dead load and the construction and traffic loads.

---

[36] *Yielding* indicates that the stress in the material is above its *yield stress*, at which level a material becomes more compliant, and irreversible deformation occurs. Below yield stress, deformations are reversible.

[37] The models showed that, at the time of the accident, the load in diagonal member L9/U10W was slightly above its design load, but the stress associated with this load was still well below the yield stress.

[38] An instability is evident in the model when it cannot balance increasing applied loads. The model showed that, at the point of instability, the highly stressed gusset plates at the U10W node are unable to resist the lateral shifting of the upper end of diagonal L9/U10. At that point, the lateral shift of the upper end of the diagonal will proceed even with no added load.

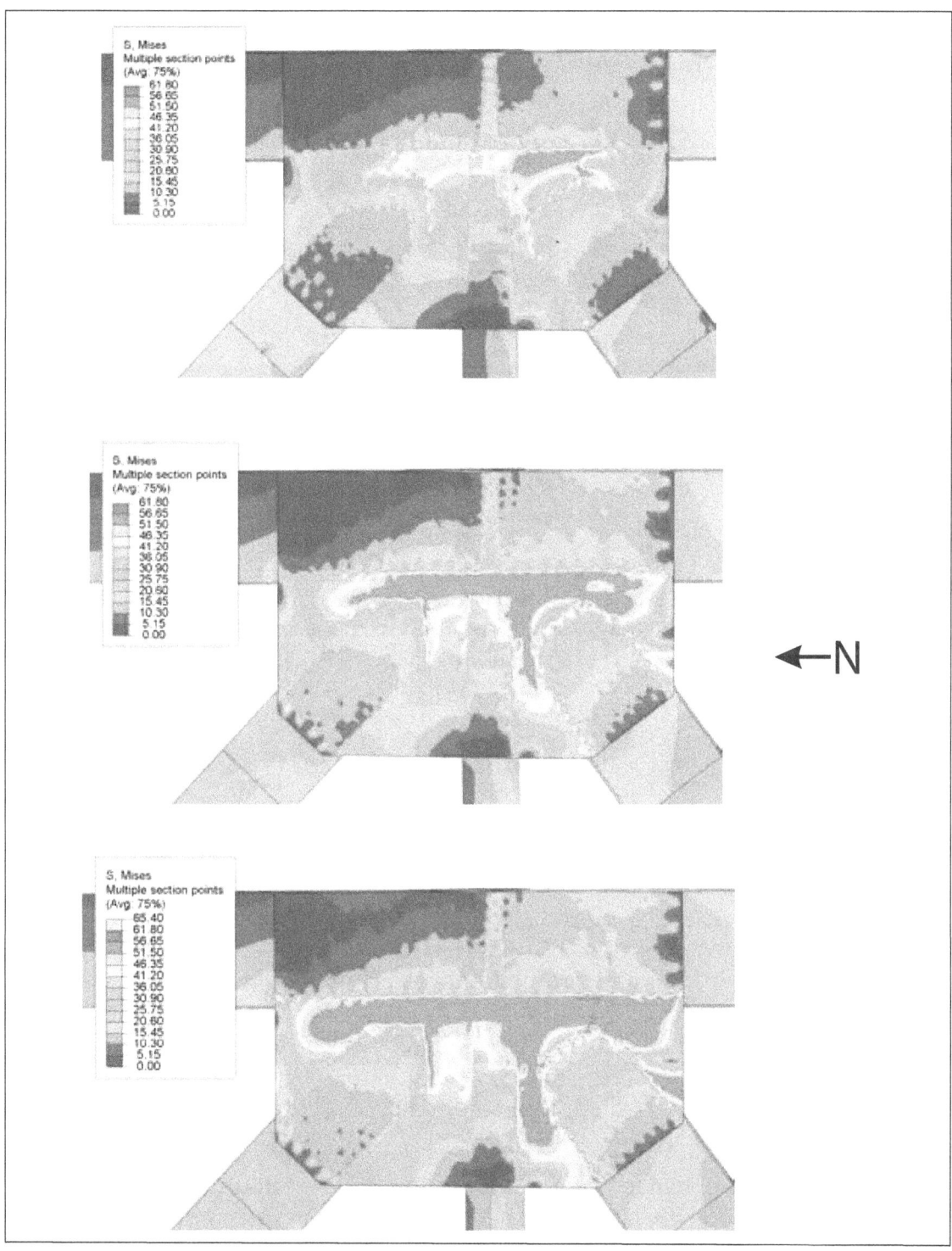

Figure 21. Finite element model stress contours for outside (west) gusset plate at U10W at time of bridge opening in 1967 (top); after 1977 and 1998 renovation projects, which increased deck thickness and modified barriers (middle); and on August 1, 2007 (bottom). Note areas that are yielding (dark orange and red contours above yield stress of 51.5 ksi) around ends of diagonals, beginning with original bridge and becoming larger as loads were increased.

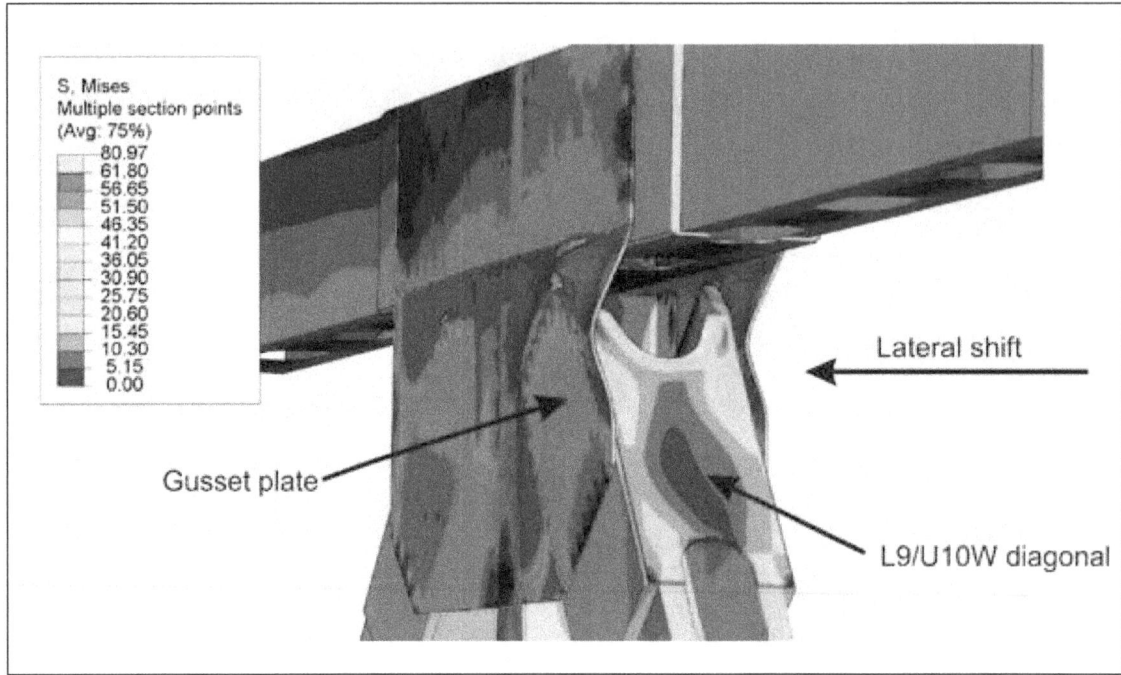

Figure 22. View of node U10W looking north, indicating lateral shift west of upper end of L9/U10W diagonal member at point of instability. (For purpose of illustration, the amount of lateral displacement, including bowing of gusset plates, is exaggerated by a multiple of 5.)

The model showed that when the U10 node gusset plates were bowed to the west, as depicted in the photographs, the upper end of diagonal L9/U10W shifted laterally to the west at the onset of instability, consistent with what was observed in bridge components after the accident. Another model of U10W, with initially flat gusset plates, also predicted an unstable lateral shift of the upper end of diagonal L9/U10W, but this shift was toward the east—the opposite of what was observed in postaccident examination of bridge components. In addition, compared to the models with bowed gusset plates, the model with initially flat gusset plates required a larger total load to trigger the lateral shifting instability.

Detailed models of both U10W and U10E were built into the global model to explore how load was shared between the two nodes. These detailed models included bowed gusset plates (all bowed to the west, as shown in the photographs), with the same initial maximum deflection at all four gusset plates. Although the bridge dead loads were symmetric, the construction vehicles and materials were located in the southbound lanes on the west side of the bridge, between piers 6 and 7, at the time of the collapse. The models showed that, under the bridge dead load and the traffic and construction live loads, the stresses in the U10W gusset plates were significantly higher than the stresses in the U10E gusset plates. The models also indicated that diagonal L9/U10W would be the first to experience a lateral shifting instability.

The effects of changes in temperature at the U10 nodes were also studied with the models. Data from a weather station at the University of Minnesota showed that, on the day of the collapse, the temperature increased from a low of 73.5° F earlier in the day to 92.1° F at 6:01 p.m. Thermal expansion should be accommodated by the movement of roller bearings; and if the roller bearings of the bridge were moving freely, such a temperature change would not affect stresses in the bridge members or gusset plates. However, measurements of roller bearing performance made by URS in 2006 suggested that the bearings moved under seasonal temperature changes but resisted motion in the short term. In the FHWA global model, the bearings were assumed to be fixed (but with some flexibility in the piers), so that a temperature change would affect the stresses. The models showed that an increase in temperature reduced the stress in the U10 gusset plates, and thus a slightly higher applied load was required to trigger the lateral shifting instability of the upper end of diagonal L9/U10W.

The FHWA investigated an additional case allowing for a difference in temperature on the east and west trusses as a result of solar radiation. Data from a study following the 1996 gusset plate failure and near collapse of the eastbound Lake County deck truss bridge over the Grand River in Ohio were used to estimate the temperature difference between the east and west trusses of the I-35W bridge on August 1, 2007. By the time of the collapse, about 6:00 pm, it was estimated that the temperature of the west truss was about 11.5° F higher than the ambient temperature and that the east truss was about 1.5° F higher. These temperature increments were added to the lower chord, diagonal, and vertical members of the two trusses; the shaded upper chord and the deck were assumed to be at ambient temperature. When compared to a case with a uniform temperature change, the effects of the temperature differentials between the two trusses were minimal, with the member forces at the U10W and U10E nodes changing by less than 2 percent in the chords and diagonals and by less than 5 percent in the verticals.

Both the FHWA and SUNY/Simulia models were also used to evaluate the effect of corrosion in the L11 node gusset plates. The corroded condition of these gusset plates was modeled as a local thickness reduction of 0.1 inch (section loss of 20 percent) running along the top of the lower chord members. The models showed that the maximum stress in the gusset plates of the L11 nodes occurred in the area of corrosion at the lower end of tension diagonal U10/L11W, but that the stresses in the U10W gusset plates were substantially higher than the stresses in the corroded L11W gusset plates under the conditions at the time of the collapse. The models also predicted that the corroded L11 nodes would support much higher loads than those necessary to trigger the instability at U10W.

The FHWA global model was also used to quantify the loads in the five main truss members that connect at the U10W node. The quantified loads were those from the original design, following the 1977 and 1998 bridge modifications, and those on the day of the accident. Figure 23 and table 10 show the fraction of the design load in each member for each condition analyzed. The design load and type of load for each member are shown in table 11. The design load includes the dead load of the original bridge design plus a maximum live load and impact load calculated per AASHO specifications. The intent of the design methodology was to ensure that, under the design load, the stress in each member would be less than the allowable stress for that member. This allowable stress would have been no more than 55 percent of the yield stress, depending on the type of load applied (tension, compression, or shear). These member design loads would also have been used to design the U10 gusset plates to satisfy the allowable stress requirements per the AASHO specifications.

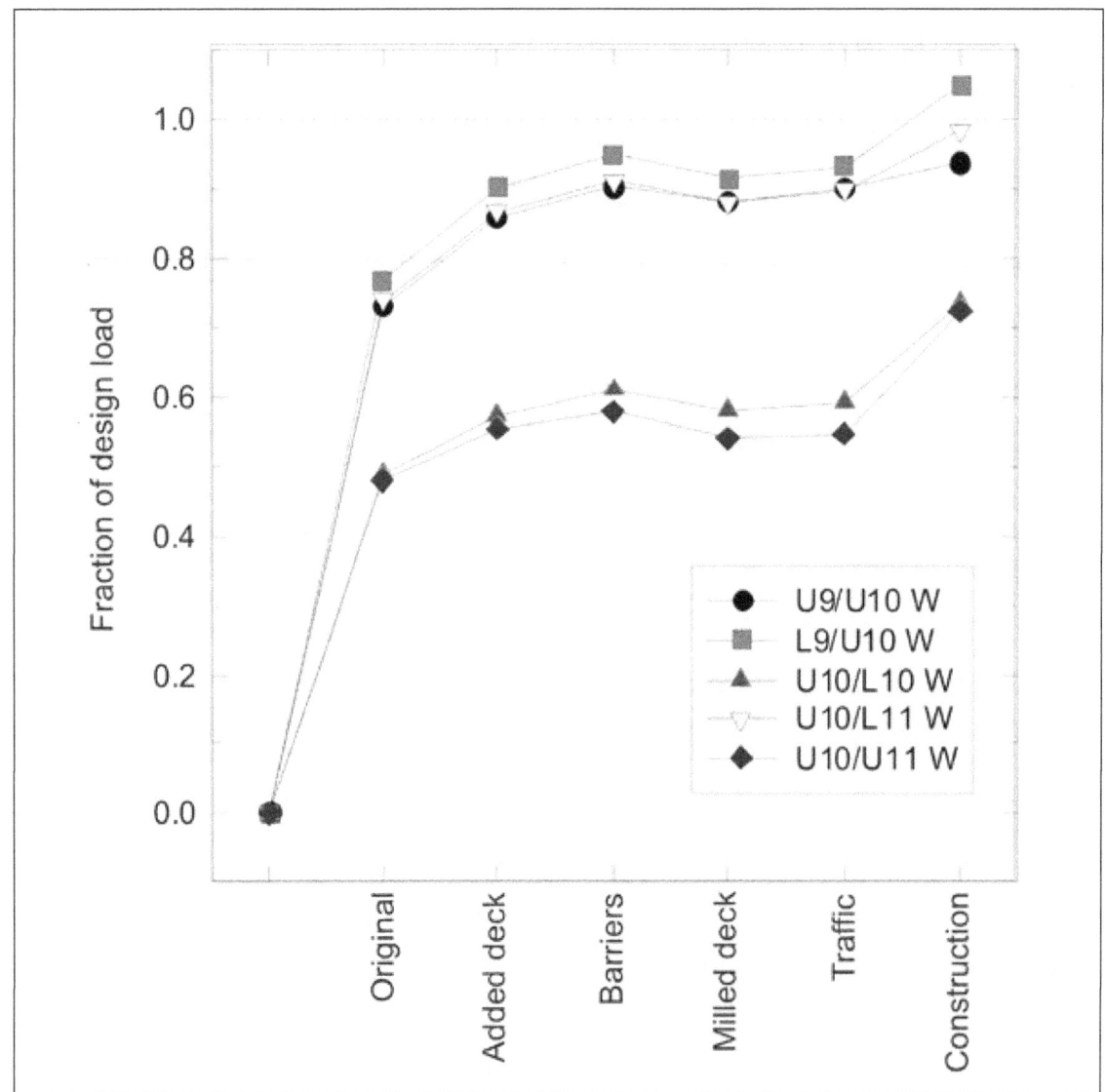

Figure 23. Loads (as fraction of design load) in five main truss members that connect at U10W node, as calculated using FHWA global model. Steps include dead load of original design, added deck thickness in 1977, and modification to barriers in 1998. The final three steps include conditions on the day of the accident: milled-off deck thickness in southbound lanes, traffic at collapse, and construction materials and vehicles.

Table 10. Fractions of design loads in five main truss members at U10W over bridge life.

	U9/U10W	L9/U10W	U10/L10W	U10/L11W	U10/U11W
Step	**Fraction of design load**				
Dead load of original design	0.73	0.77	0.49	0.74	0.48
Added deck in 1977	0.86	0.90	0.57	0.87	0.55
Modified barriers in 1998	0.90	0.95	0.61	0.91	0.58
Milled-off deck in southbound lanes	0.88	0.91	0.58	0.88	0.54
Traffic at collapse	0.90	0.93	0.59	0.90	0.55
Construction loads and vehicles	0.94	1.05	0.73	0.98	0.72

Table 11. Design loads for five main truss members at U10W (or U10E) node.

	U9/U10	L9/U10	U10/L10	U10/L11	U10/U11
Design load (thousands of pounds)	2,147	2,288	540	1,975	924
Mode	Tension	Compression	Tension	Tension	Compression

At each load step, the most highly loaded member was L9/U10W, both in absolute terms and as a fraction of the member design load. At the final step in the calculation, the load in that member exceeded the design load by about 5 percent. With regard to the load in this member at the time of the accident, the dead load of the original bridge design contributed about 73 percent, the added deck thickness contributed about 13 percent, the modified barriers contributed about 5 percent, and the construction materials and vehicles contributed about 11 percent. The milled-off southbound lanes reduced the load by about 3 percent, and the traffic added back about 2 percent.

Design History of I-35W Bridge

The I-35W bridge design process started on October 22, 1962, when the State of Minnesota entered into an agreement with Sverdrup & Parcel to produce (1) a preliminary engineering report, (2) final design plans (checked by a registered professional engineer), and (3) checked design computations for a new interstate bridge, number 9340, across the Mississippi River. With regard to the materials of construction, the agreement stated, "The use of steels of various strengths shall be investigated to determine the advisability for use either in whole or in part for stress carrying members." The agreement showed that Sverdrup & Parcel's responsibilities did not include checking shop detail drawings for fabrication.

Evolution of Design

Safety Board investigators examined documents related to design of the I-35W bridge, including design studies, engineering drawings, construction plans, and both interagency and intra-agency correspondence and notes that were provided by Mn/DOT and Jacobs Engineering. This document review revealed that a progression of conferences, plan reviews, and design revisions took place with regard to types of steel and related construction details in the main trusses of the bridge[39] from the time of the design consultant's preliminary design report in April 1963 until the final plans were certified by Sverdrup & Parcel in March 1965 and subsequently approved by Mn/DOT in June 1965. This section summarizes those reviews and design changes.

At the time the I-35W bridge was built, the highway authority in Minnesota was referred to as the Minnesota Department of Highways or, at times, the Minnesota Highway Department. The highway authority has evolved into Mn/DOT, and the latter designation is used in this section. The Federal authority at the time of bridge construction was the Bureau of Public Roads within the U.S. Department of Commerce. That organization has evolved into the FHWA, which is used in this section.

Table 12 lists the four types of steel, referred to by their ASTM[40] equivalents, that were recorded in various design documents and were eventually used in the I-35W bridge.

Table 12. Types of steel used in I-35W bridge.

ASTM equivalent	Specified minimum yield stress (psi)[A]	Allowable stress (psi)[B]
A36	36,000	20,000
A242 (plates 3/4 to 2.5 inches thick)	50,000	27,000
A441 (plates less than 3/4 inch thick)	50,000	27,000
A514 and T-1[C]	100,000	45,000

[A] *Yield* is the stress at which strain in a material changes from elastic deformation to plastic (irreversible) deformation.

[B] *Allowable stress* is the maximum stress, as specified in design criteria, that should be experienced by any bridge component.

[C] T-1 was originally a U.S. Steel trade name for a structural steel meeting ASTM specification A514.

---

[39] This examination did not address documentation or issues related to the floor trusses or the approach spans.

[40] ASTM International, formerly known as the American Society for Testing and Materials, is an international standards organization that develops and publishes standardized testing methods to evaluate a range of materials, products, systems, and services.

Initial Design and Preliminary Engineering Report, October 1962–April 1963

The preliminary engineering report presented to Mn/DOT on April 12, 1963, contained both a two-truss design and a four-truss design but recommended the two-truss design for economic reasons. Mn/DOT selected the two-truss design. The preliminary engineering report was primarily narrative and included no design drawings.

With regard to the steel to be used, the report stated, "Welding is planned throughout for the make-up of girders and truss members and, in this connection, high yield strength steel conforming to . . . T-1 . . . will be used extensively." The report noted that "an all welded structure would require approximately 20% less steel than a riveted structure, with a possible resultant cost saving of more than 10%."[41] The report summarized material weights but did not provide sizes and materials for any members or gusset plates.

In support of the preliminary engineering report, Sverdrup & Parcel had generated an initial internal two-truss design, which was documented by a set of unchecked deck truss design computations with sheets dated from December 19, 1962–January 21, 1963.[42] These computation sheets provided the first indication of the materials intended for use in construction of the deck truss spans of the bridge. Among other calculations, the computation sheets detailed the D/C calculations for members from nodes 0–14. The capacity calculations provided the initial member side plate thicknesses and materials. For the chord and diagonal members, 15 were specified as T-1 steel, 11 as A441 steel, and 4 as A36 steel. The vertical members were all listed on the computation pages as A36 steel. No gusset plate materials were specified in this initial design.

By September 1963, both Mn/DOT and the FHWA began to express reservations about using such large quantities of T-1 steel in the bridge. A primary concern within Mn/DOT was that the individual T-1 members would require splices because they were shorter.[43] Also, using the longer A441 members would allow some upper and lower chord members to be continuous through some nodes, greatly simplifying those connections.[44] T-1 members would not be continuous through any node, necessitating more complex joints at all node locations.

---

[41] According to the preliminary report, the cost of T-1 steel was $0.38 per pound, and the cost of A441 steel was $0.31 per pound.

[42] The computation pages were marked 1–25 but actually numbered 31, with several subpages inserted into the number sequence.

[43] The length of T-1 members is limited by the size of the heat treatment facilities where they are produced. A441 and A242 members are heat treated in a continuous process, with much greater lengths available.

[44] In the final design, the upper chord members between even-numbered nodes and the lower chord members between odd-numbered nodes were continuous beams at least 72 feet long. Thus, the odd-numbered nodes on the upper chord and the even-numbered nodes on the lower chord had simplified joints, with the verticals attached to the continuous upper or lower chord.

On December 4, 1963, Mn/DOT asked that Sverdrup & Parcel provide drawings showing a typical connection (node) layout because none of the drawings and documents that had been provided up to that time had detailed node design features or indicated material selections for specific locations. On December 13, 1963, the designer provided Mn/DOT with a detail drawing of the U12 node.

The U12 detail drawing showed T-1 steel being used in two of the five structural members of the node as well as in the 0.5-inch-thick gusset plates, the joint splice plates, and the lightly loaded lateral attachment plates and angles. In a February 14, 1964, letter to Mn/DOT about the U12 joint details, the FHWA stated:

> The proposed truss joint detail at U12 appears to be satisfactory, except [that the] unbalanced rivet pattern and resulting eccentric connection of the truss diagonals is considered undesirable and unnecessary. It is recommended that the gusset plates be enlarged to facilitate a balanced connection.

This statement appears to have been a reference to the asymmetric geometries and rivet patterns at the upper ends of the diagonals L11/U12 and U12/L13. In the U12 drawings, the upper corners of both diagonals were heavily chamfered (clipped), while the lower corners remained relatively square.

Preliminary Design, April 1963–March 1964

The Safety Board closely reviewed all available design documents for information related to gusset plate calculations. The only document containing any calculations related to the gusset plate materials and thickness was a set of unchecked computation sheets dated from November 1963–January 1964. These sheets contained information on member materials and some gusset materials used in preliminary design of the deck truss. The computation sheets showed calculations to determine the number of rivets for each node of the truss. Additional calculations used estimates of the load carried across the chord splices (upper or lower chord) to determine the thickness of the gusset plates. These calculations only considered forces carried by the upper and lower chord members and neglected any forces associated with the diagonal and vertical members.

Correspondence indicates that Sverdrup & Parcel presented its preliminary design for the bridge to Mn/DOT and the FHWA at a conference among the parties in March 1964. Although the materials of many members changed between the initial and preliminary designs, about half of the chords and diagonals remained specified as T-1 steel in the preliminary design.

At the March 1964 conference, Mn/DOT and the FHWA directed that Sverdrup & Parcel eliminate T-1 steel from all main truss members and use A441 and A242 steels instead. They also directed that the ends of the diagonals at connections (as represented by drawings of the U12 node) be chamfered symmetrically. The arrangement of openings in the member cover plates (required for internal inspection or fabrication) was discussed, and the final decision was that the cover plates of all of the box members would be perforated except for the top cover plates of the upper chord members.

The member materials listed in the computation pages match those indicated in the annotated drawing of the preliminary design, which also included materials for several vertical members not previously identified. Additionally, the gusset plates for upper and lower nodes 0–14 were identified for material and/or thickness. The gusset plates at both the U10 and L11 nodes were shown as 0.5-inch A441 steel, and they did not change from the preliminary design to the final design. In comparison, the gusset plates at the U12 node were identified as 0.5-inch T-1 steel in the computation sheets and in the U12 detail provided to Mn/DOT but were changed to 1-inch-thick A441 steel for the final design.

Final Design, March 1964–March 1965

Because of the decision regarding T-1 steel and the lower allowable stress levels for the required A242/A441 steels (27,000 psi for A242/A441 steels versus 45,000 psi for T-1), Sverdrup & Parcel had to redesign all the truss members and gusset plates for which T-1 steel had initially been proposed. For example, the thickness of all tension members would have to be increased by about 67 percent to maintain the same tensile stress capability. These thicker members increased the dead load on the bridge, which required additional design computations. The use of thicker steel did allow many of the members to also be reconfigured into simple box members without a centerline web, thereby facilitating fabrication and offsetting some of the increased weight.

Safety Board investigators found no apparent additional changes in material or material specification between the March 1964 conference and the final design plans certified by Sverdrup & Parcel on March 4, 1965, except for changes related to adopting the Mn/DOT steel specification numbering system for the final plans. Further, review of the Allied Structural Steel Company's shop plans, used to make individual components for the bridge, revealed no changes in the materials or thicknesses of the U10 or L11 gusset plates from the final design plans. Physical testing of samples from each gusset plate from the U10(') nodes confirmed that the steel met the minimum specified yield stress requirements.

Other Information

Mn/DOT Postaccident Actions

Independent Review of Designs for Major Bridges. Since the accident, Mn/DOT has revised its LRFD Bridge Design Manual to require that—for all major bridges[45] designed by consultants—an independent review be conducted by a second design firm. The stated purpose of this requirement is "specifically to reduce the potential for a design error in the contract plans." The designs of bridges considered to be "routine" will continue to be reviewed by in-house Mn/DOT staff.

Gusset Plate Reviews. According to Mn/DOT, by the fall of 2007, the agency had developed a procedure for performing engineering reviews of gusset plates on the 25 truss bridges in the State system. Reviews of several trusses were underway when Safety Recommendation H-08-1 and the FHWA's Technical Advisory T 5140.29 were issued in January 2008. Mn/DOT subsequently retained consultants to perform similar reviews of bridges in the county and township systems. The reviews consisted of performing a complete load rating of the trusses; performing a design check of the gusset plates using loads from the ratings as well as inspection information; and, for some bridges, conducting an additional field review to supplement previous inspections.

Bridge Inspections. Mn/DOT officials reported having completed an accelerated schedule of inspections of all State bridges by December 2007, with the information from those inspections to be used in planning for future maintenance. Based on these inspections, one bridge was closed temporarily in August 2007 for steel repairs, and another was closed briefly to repair damage caused by a vehicle. Mn/DOT also contracted with a consultant to assess its compliance with the National Bridge Inspection Standards.

Documentation of Postinspection Bridge Maintenance Decisions. Since the collapse of the I-35W bridge, Mn/DOT has developed standard practices for documenting postinspection bridge maintenance decisions. A consultant was also retained to assist the agency in a quality improvement review of transportation district and bridge office procedures.

Bridge Maintenance and Inspection Staffing. Mn/DOT has begun an assessment of the sufficiency of bridge maintenance staffing at the district level. Staffing for fracture-critical inspections was increased, and the need for an additional snooper vehicle was identified.

---

[45] Mn/DOT defines "major bridges" as bridges containing spans 250 feet and greater in length. Additionally, the bridge design engineer may elect to require peer review for unique bridge types. An exception to this requirement is steel plate girder bridges, reviews of which will continue to be performed by in-house design units.

Construction Loads on Bridges. After the collapse, Mn/DOT revised its policies regarding the placement of construction loads on bridges, as discussed elsewhere in this report.

Deficient Gusset Plates on Other Minnesota Bridges

During the postcollapse engineering reviews of gusset plates that Mn/DOT conducted on the remaining 25 steel truss bridges in the State system, a number of deficiencies were found, as summarized below.

DeSoto Bridge. On March 20, 2008, Mn/DOT permanently closed the DeSoto Bridge in St. Cloud, Minnesota (built in 1957), after "bending" of the gusset plates was identified at four locations on the lower chord of the main truss. Because of distortions in the L11 and U6 gusset plates, Mn/DOT contracted with Wiss, Janney, Elstner Associates, Inc., to evaluate the bridge. The consultants reviewed the original Sverdrup & Parcel design drawings, the Illinois Steel Bridge shop drawings, the 1978 deck repair and overlay plans, and various inspection reports. The consultants determined that the distortion in the gusset plates had occurred during the original fit-up because of imperfect match of member depth, misalignment of members, or typical erection stresses. No signs of significant distress or deterioration in the gusset plates were found, and it was determined that the plates were adequately proportioned and that their load-carrying capability had not been compromised by the bending. The L11 and U6 gusset plates did exceed the standards for length of unsupported edge.

The most recent routine inspection of the DeSoto bridge was in August 2007, and a fracture-critical inspection was performed in September 2007. Mn/DOT officials advanced the scheduled replacement date for the bridge from 2015 to 2008. According to Mn/DOT, work began on August 25, 2008, to remove the old bridge to prepare for the new construction. The new bridge is expected to be in place by November 2009.

Blatnik Bridge. The John A. Blatnik bridge was opened in 1961 and carries Interstate 535 across the St. Louis River between Duluth, Minnesota, and Superior, Wisconsin. Mn/DOT closed two of the four lanes on the bridge on May 6, 2008, after 16 gusset plates were found to have less than the desired or expected safety factor. The two lanes were closed to reduce the load on the bridge and to allow room for contractors to add additional steel to bolster the strength of the 16 plates. According to Mn/DOT, the original design of the plates was adequate, but their safety factor was compromised in the early 1990s by the addition of 2 inches of concrete to the bridge deck, which had increased the dead loads on the structure. From June 9–21, 2008, workers installed steel angles to strengthen the deficient gusset plates. The Blatnik bridge had had a fracture-critical inspection in August 2007 and a routine inspection in October 2007.

Highway 43 Bridge. This bridge, built in 1941 and remodeled in 1985, carries Highway 43 over the Mississippi River in Winona, Minnesota. On June 3, 2008, Mn/DOT closed the bridge because of rust and corrosion on gusset plates at several locations, some of which had penetrated the metal. Some bulging of plates was also found at one location. Ultrasonic testing of some of the plates revealed loss of plate thickness ranging from 25–100 percent (complete penetration) of the nominal plate thickness. Repairs to the corroded plates were completed on June 21, 2008.

From July 30–to August 1, 2007, the Highway 43 bridge had undergone a fracture-critical inspection. The report of that inspection indicated that some gusset plates had extensive pitting (some of which had been painted over), with some penetration of the plates. One gusset plate was reported to have 60 percent section loss along a bottom chord connection. The report noted, "There are cracked tack welds along some of the deck truss bottom chord field splices . . . these should be monitored during future inspections."

The inspection report did not suggest actions or monitoring with regard to the rust and corrosion found on the gusset plates. The "Summary and Recommendations" section stated, in part,

> No critical structural deficiencies were observed during this inspection.
>
> The truss members and connections located at or below the deck level have paint failure and corrosion (pack rust and section loss). While the section loss is not yet severe enough to require repair or load restrictions, the truss spans (Spans #16-24) should be repainted to prevent further section loss due to corrosion.

The Highway 43 bridge also had a routine inspection on April 16, 2007. With regard to section loss, that inspection report stated, "There are scattered areas on most steel members with minor loss of section that has been cleaned and painted over or that occurred after the last paint contract." Regarding pack rust, the report stated:

> Most steel members with faying surfaces have pack rust showing with some areas having rust between the plates that has caused serious distress at the connection, however all connections are still functioning.

The April 16, 2007, report noted no "critical findings" for the Highway 43 bridge.

Highway 61 Bridge. This bridge, built in 1950, carries Highway 61 over the Mississippi River in Hastings, Minnesota. It is a steel high-through-truss bridge with 3 main spans and 10 approach spans. An April 2008 bridge inspection found a number of lower truss gusset plates that required additional bracing because of distortion (generally bowing) in the plates. In response to the Safety Board's preliminary findings with regard to the I-35W bridge, Mn/DOT retained a consultant to provide a followup review of the Highway 61 bridge and to evaluate the pack rust that had been found in the lower panel gusset plates. The consultant attributed the gusset plate distortion (maximum 7/16 inch) primarily to the pack rust, stating, "Gusset plate pack rust and their slight distortion is a common observation found in riveted bridges of this era and historically has not been cause of alarm." The gusset plates were strengthened under a previously scheduled bridge painting and preservation contract, during which workers identified an additional gusset plate that required strengthening because of corrosion. The affected gusset plates were repaired. Replacement of this Structurally Deficient bridge[46] has been accelerated to 2010.

Deficient Gusset Plates on Ohio Bridges

During the investigation of the I-35W bridge collapse, Safety Board investigators reviewed the circumstances surrounding two steel deck truss bridges in Ohio in which either the gusset plates had failed or their conditions had raised concerns about structural integrity.[47] In 1996, gusset plates in the superstructure of the eastbound Lake County bridge over the Grand River failed while maintenance crews were working on the structure. In 2007, emergency repairs had to be undertaken to address the deterioration and deformation of gusset plates in the superstructure of the Innerbelt bridge over the Cuyahoga River Valley.

Lake County Grand River Bridges. The Lake County Grand River bridges were twin structures spanning the Grand River in Lake County, Ohio, about 30 miles east of Cleveland. Each structure comprised five spans totaling 863 feet. Spans 1 and 5 were 75-foot-long simply supported approach spans. Spans 2, 3, and 4 were approximately 208, 297, and 208 feet long, respectively. Spans 2 and 4 were arched, cantilevered deck trusses supporting a suspended truss section in span 3. Each bridge had a deck width of 44 feet and carried two traffic lanes of Interstate 90. (One bridge carried eastbound traffic, the other westbound.) The bridges were designed in 1958 by Capitol Engineering Associates of Dillsburg, Pennsylvania, and opened to traffic in 1960.

On May 24, 1996, the superstructure of the eastbound bridge was being painted, and a temporary work zone had been established by closing the right traffic lane and shoulder. Vehicles and equipment related to the painting project occupied the right shoulder in the area over the L8' node. The left lane remained open to traffic.

---

[46] This bridge was also *Functionally Obsolete* because of inadequate clearances, but the National Bridge Inspection Standards provide for only a single condition rating for each bridge.

[47] The information in this section is based on documentation provided by the Ohio Department of Transportation (ODOT).

As a truck traveled over the bridge in the left lane, all four of the 7/16-inch-thick gusset plates at the L8' nodes buckled, causing a 3-inch lateral displacement. This displacement, in turn, allowed the compression members at the L8' connections to move downward about 3 inches. Following the failure, ODOT closed the bridge to traffic and initiated an investigation.

The investigation identified the source of the failure as corrosion of the L8' gusset plates, which had resulted in significant section loss and had penetrated completely through the plates at some locations. It was determined that a leaking joint above panel point U9' had allowed salt-contaminated water to run down diagonal U9'-L8' to the lower chord gusset plates. The years of corrosive runoff had resulted in crevice corrosion, the byproduct of which had manifested itself as thin sheets of layered rust. Additionally, oxygen-rich corrosion cells had started to take root along the vertical faces at the insides of the gusset plates. The result was a line of section loss that rendered the gusset plates incapable of handling the additional loads created by the maintenance project on the day of the incident. The investigation revealed that the blooms of oxidation concealed perforations in the base metal such that the extent of corrosion could not be, and had not been, adequately assessed through visual bridge inspections.

After the failure in the eastbound Lake County Grand River bridge, ODOT sought and received the assistance of the FHWA in performing a failure analysis of the structure. ODOT also contracted with Richland Engineering Ltd. to conduct an independent analysis of the gusset plates on the two Grand River bridges. These analyses revealed that many of the gusset plates did not meet code requirements with regard to unsupported edge lengths. Engineers from Richland used several methods to assess the gusset plates, including some based on buckling criteria. They highlighted several gusset plates as being somewhat underdesigned, with the L8 and L8' nodes identified as the most underdesigned using one of the buckling criteria, even when assuming that the gusset plates had no corrosion. Representatives of Richland and ODOT told the Safety Board that, though they believed some gusset plate design elements contributed to the failure, they considered the primary cause to be corrosion of the plates.

In the wake of the failure, ODOT initiated emergency repairs to the four corroded L8' gusset plates on the eastbound bridge and added stiffeners to 46 of the gusset plates on each bridge structure. The four L8' gusset plates on the eastbound structure were the only ones that required the replacement of structural material.

FHWA representatives told the Safety Board that the agency took no followup action and issued no advisories as a result of the gusset plate failure on the Lake County Grand River bridge. They said that the failure was attributed to section loss from corrosion and that bridge inspection standards were already in place to require that structures be examined for this condition.

Cuyahoga County Innerbelt Bridge. The Innerbelt bridge spanned the Cuyahoga River Valley on the north side of Cleveland and carried eight lanes of Interstate 90 traffic through the downtown area. Including the approach spans, the bridge was 5,079 feet long. The main truss portion of the bridge consisted of nine cantilevered, arched deck truss spans supporting a reinforced concrete deck with steel curbs, safety walks and railings, and a concrete median barrier. The total length of the truss spans was 2,722 feet. The bridge was designed in 1955 by Howard, Needles, Tammen & Bergendoff, of Cleveland, Ohio, and was opened to traffic on August 15, 1959.

An inspection of the bridge in October 2007 revealed that the inside and outside gusset plates at many connections exhibited crevice corrosion along the vertical face of the plates and along the top of the lower chord. As a followup to this inspection, ODOT inspectors performed an additional evaluation of the bridge using nondestructive evaluation methods. All 468 gusset plates on the bridge were measured for section loss using hand-held ultrasonic thickness gauges. These measurements indicated that visual inspections of the plates had significantly underestimated the amount of section loss. At some locations, the corrosion was found to be accompanied by deformation. The magnitude of the deformation, or bowing, of the gusset plates had been documented by placing a straightedge along the plates at various locations and measuring the gaps between the straightedge and the plates.

Based on its nondestructive evaluation examination of the Innerbelt bridge, ODOT initiated an emergency repair program, which it completed in April 2008. For the gusset plates having the most corrosion and the greatest amount of section loss, the repairs involved bolting an additional plate to the outside of the original gusset plate and adding stiffeners. In all, the gusset plates at 21 locations were repaired in this manner. The plates at an additional 12 locations were repaired using stiffening angles only. ODOT officials told the Safety Board that the agency plans to continue addressing the gusset plate problems on the Innerbelt bridge and that additional gusset plates will be strengthened during a project planned for later in 2008.

ODOT Bridge Inspection Program. ODOT adheres to the National Bridge Inspection Standards as set forth by the FHWA in the National Bridge Inspection Program. Bridge inspections in Ohio are conducted by private consultants as well as ODOT inspectors. From 1981 until 2007, inspections of the Innerbelt bridge were performed by a variety of consulting firms. An additional 2007 inspection was performed by ODOT inspectors, who are not required to be professional engineers, though ODOT officials said that the consultants hired by the State typically employ professional engineers to conduct bridge inspections.

Guidance and Training for Inspecting Gusset Plates

The Safety Board examined the current FHWA bridge inspection guidance documents, training materials, and training programs to determine the extent to which they contain information specific to the evaluation and condition rating of gusset plates on steel truss bridges. This review focused primarily on the *Bridge Inspector's Reference Manual* and the training programs of the National Highway Institute.

Bridge Inspector's Reference Manual. The *Bridge Inspector's Reference Manual* provides bridge inspectors with information and guidance regarding the programs, procedures, and techniques for inspecting and evaluating highway bridges. The manual is divided into 13 sections, each dealing with a specific concept or evaluation area. Each of these sections is subdivided into "topics" and subtopics that give detailed information about individual elements. The sections and topics that are most applicable to the inspection and evaluation of steel truss bridges are as follows:

- Section 2, "Bridge Materials" (Topic 2.3, "Steel"),

- Section 8, "Inspection and Evaluation of Common Steel Superstructures" (Topic 8.1, "Fatigue and Fracture in Steel Bridges"), and

- Section 8, "Inspection and Evaluation of Common Steel Superstructures" (Topic 8.6, "Steel Trusses").

The specific topics and subtopics included in the Safety Board's review of the *Bridge Inspector's Reference Manual* are discussed below.

Topic 2.3, "Steel." This topic area contains information pertinent to the properties of steel, including the types and causes of steel deterioration. It addresses fatigue cracking, bending/distortion, and overloads (including elastic and plastic deformation) in bridge members. Floor beams are the only types of bridge members specifically identified, with general references to tension or compression members. Gusset plates are not mentioned.

Topic 8.1.2, "Failure Mechanics." This topic area addresses the various types of fractures that can occur in steel, including plastic deformation in a member. The information is presented in general terms that would direct an inspector to examine any applicable deformation, but it does not specifically refer to gusset plates as a possibly affected member.

Topic 8.1.8, "Inspection Procedures and Locations." This topic area discusses issues specific to a fracture-critical member, including gusset plates at lateral bracing locations, where cracking or out-of-plane distortion should be noted and evaluated. But the members referenced do not include gusset plates at panel points along the main chords.

Topic 8.6.4, "Inspection Procedures and Locations." This topic area discusses issues related to truss bridges, including components such as tension members, compression members, and floor systems. As with other sections in the manual, the presentation of material is designed to provide an inspector with a general knowledge of conditions such as corrosion (section loss) and buckling, as well as where and how these conditions may be detected. Attention is given to the primary members and to specific locations, such as the floor system, but nothing in the material specifically addresses gusset plates. A discussion of secondary members addresses the propensity of horizontal plates to corrode. Additionally, information is provided about other areas that are particularly susceptible to corrosion because of trapped water and debris, such as "tightly packed panel points," which would be applicable to gusset plates, but there is no guidance as to what constitutes such a condition within a gusset connection.

National Highway Institute Training Courses. The National Highway Institute was established by Congress in 1970 to provide surface transportation training, resource materials, and educational opportunities. The institute provides a 3-week training program for bridge inspectors that comprises a 1-week course, "Engineering Concepts for Bridge Inspectors," and a 2-week course, "Safety Inspection of In-Service Bridges." When combined, these courses, which are based on the Bridge Inspector's Reference Manual, meet the FHWA requirements for a comprehensive training program in bridge inspection as defined in the National Bridge Inspection Standards.

Safety Board investigators examined the *Instructor's Guide* and the *Participant's Workbook* for the 2-week "Safety Inspection of In-Service Bridges" course for specific references to gusset plate corrosion or deformation. The guidance presented in this material was the same as that found in the *Bridge Inspector's Reference Manual*.

Another National Highway Institute course is "Fracture Critical Inspection Techniques for Steel Bridges." This 3.5-day course includes training and hands-on workshops for nondestructive testing equipment as well as a case study on preparing an inspection plan for a fracture-critical bridge. The curriculum covers inspection procedures and reporting for common fracture-critical members, such as problematic details, I-girders, floor beams, trusses, box girders, pin and hanger assemblies, arch ties, eyebars, and cross girders/pier caps.

Safety Board investigators reviewed the *Instructor's Guide* and the *Participant's Workbook* for this training course for issues specific to gusset plates. Each set of training materials contained general references to gusset plates. The most significant finding for plates at panel points was found under Session 4, Topic 3, "Inspection Procedures: Trusses." This topic area contained the following paragraph:

> Gusset Plates - The other mechanical connection common on truss members is their connection at the **truss panel points**. If not pin connected, **gusset plates** are used to transfer loads at panel points. A positive connection can be made using **rivets, bolts** or a combination of both. During inspection, the fasteners and the surrounding gusset plate area should be examined closely.

FHWA Technology Initiative. In Section 8 of the Bridge Inspector's Reference Manual, "Inspection and Evaluation of Common Steel Superstructures," Topic 8.1.8 discusses "Inspection Procedures and Locations." This section, under the subsection "Procedures," presents the following guidance:

> The inspection of steel bridge members for defects is primarily a visual activity. Most defects in steel bridges are first detected by visual inspection. In order for this to occur, a hands-on inspection, or inspection where the inspector is close enough to touch the area being inspected, is required. More exact visual observations can also be employed by cleaning suspect areas, removing paint when necessary, and using a magnifying unit.

Additional material related to inspection practices includes a list of advanced inspection techniques that may be used to find faults or deficiencies that would not always be detected through visual inspections alone. The manual does not go into detail concerning the use, benefits, or limitations of these techniques.

Following the I-35W bridge collapse, a group representing the various bridge program offices within the FHWA met and mapped out short- and long-term plans for improving the tools and approaches available to bridge owners for inspecting and assessing their highway structures, and for educating and training owners and field personnel on the availability and application of these technologies.

Among the short-term initiatives of the group was the development of the FHWA "Bridge Inspector's NDE Showcase." This 1-day showcase was developed to demonstrate five advanced bridge evaluation and inspection tools that are commercially available but may be underutilized in many State bridge inspection programs. Two of the technologies focus on steel bridge components and three on concrete bridge components. The showcase targets both the managers of bridge inspection programs and the inspectors in the field. Subjects covered include the use of each tool in the field, the type of information each provides, and their strengths and limitations for specific bridge inspection and detailed assessment situations. The five technologies presented in the Bridge Inspector's NDE Showcase are as follows:

- *Eddy current:* Uses electromagnetic induction to assess surface flaws, material thickness, and coating thickness. Typically used on metals with painted or untreated surfaces.

- *Ultrasonics:* Uses high-frequency sound energy to assess flaws (surface and subsurface) and make dimensional measurements. Typically used on metals with untreated or cleaned surfaces.

- *Infrared thermograph:* Measures the amount of infrared energy emitted by an object to calculate temperature. Typically used to assess deterioration damage, surface and subsurface flaws, and moisture intrusion.

- *Impact echo:* Uses impact-generated stress waves to assess subsurface flaws and material thickness in concrete and masonry.

- *Ground penetrating radar:* Uses electromagnetic waves to assess subsurface flaws and to image embedded reinforcement or tendons in concrete, asphalt, timber, or earthen structures.

The showcase has been pilot-tested with the New York State DOT and is currently being revised. It is expected to be available for all State DOTs by the end of 2008 or early 2009.

FHWA–AASHTO Joint Study of Gusset Plates

In May 2008, representatives of the FHWA and AASHTO proposed that the two agencies participate in a joint study of gusset plates, with the intent of developing guidance for bridge engineers in the proper design and rating of gusset connections. The proposed problem statement for the study noted that bridge connections, because they involve complex geometry and stress, present special analytical and design challenges that can lead to varying assumptions about load path and "vastly different design and rating methodologies." The statement also noted that previous gusset connection research has focused primarily on building-type bracing systems, with the few existing studies on bridge connections limited to the seismic behavior of those connections. Thus, according to the problem statement, "there exists a need for a comprehensive yet focused investigation on bridge type connections."

The main research objectives of the 24-month joint study project are as follows:

- Perform advanced finite element analyses of varying bridge gusset connection types, configurations, loadings, and failure modes to verify or modify existing procedures, or to develop new design and rating procedures.

- Perform large-scale experimental investigations to validate the findings of the finite element analyses.

- Based on the analytical and experimental investigations, develop recommendations for optimal connection configurations to maximize the resistance of gusset connections and minimize the possibility of unfavorable failure modes.

- Develop guidelines, specifications, and examples for the load and resistance factor design and rating of gusset connections.

Bridge Design Firm Process for Design and Design Review

Sverdrup & Parcel Documents and Procedures. Jacobs Engineering was unable to locate the original Sverdrup & Parcel calculations that had been used in designing the main truss gusset plates on the I-35W bridge. Nor did Mn/DOT, in its bridge construction files, have the calculations for the main truss gusset plates. Calculations for design of the floor truss gusset plates were available from both Jacobs and Mn/DOT; these calculations were made using the design methodologies commonly in use at the time and resulted in gusset plates that were properly designed for their respective loads.

Jacobs Engineering provided the Safety Board with a copy of Sverdrup & Parcel's *Procedure for Checking Design Notes and Coordinating Same with Detail Checker*, dated September 1953. This procedure specified how designs would be checked and rechecked and would have been the same or similar to the procedure used during design of the I-35W bridge.[48] The following excerpts are taken from the Sverdrup & Parcel document:

> The design notes shall be checked on the original computation sheets, not prints. The checker shall make all check marks and all change notations in blue pencil. . . .

> When the original designer is returned the checked design sheets, he shall backcheck the checker's work. This in effect amounts to the original designer checking the checker. Any disagreement with the checker's blue marks shall be noted by the designer in green pencil on the computation sheets. In checking any new sheets of computations that the checker has added to the computations the designer shall use a blue pencil since he is then acting as a design checker.

> When the design checker receives the design from the designer after it is backchecked he shall see that all his blue marks have been agreed to and the corrections made. . . . Where green marks [indicating disagreement] occur he shall see that a final figure is put in the original space and the green marks removed. . . .

> Since the checking of detail drawings may have been done at anytime during the checking of the design notes it is most important that the detail checker be allowed to examine the design notes before any checking or backchecking correction marks are removed, or voided sheets of calculations are removed from the set of design notes. If this is not done, the details will either not conform to the latest design requirements, or the completed details affected by any design revision will have to be rechecked. Either result cannot be tolerated for obvious reasons.

---

[48] One or more later versions of this procedure may have been published, but Jacobs representatives could not locate a version from the 1960s.

Jacobs Engineering also provided the Safety Board with the Sverdrup & Parcel *Quality Control Coordination and Checking Procedures*, dated April 1975. The following excerpts are taken from that document:

5.1.4 CHECK, BACKCHECK, AND RECHECK

5.1.4.1 CHECK. Upon completion of the design calculations they shall be checked by an engineer technically competent for the assigned task. Because of the progressive nature of design calculations, the checker, during his design check, shall consult with the Design Engineer on any differences which are found. If agreement between the checker and the Design Engineer cannot be reached, the matter shall be resolved as outlined in the paragraph below entitled "Backcheck". In the interest of efficiency and accuracy, as few checkers as practicable shall be used in checking the design on any one project.

5.1.4.2 BACKCHECK. Upon completion of his check the checker shall return the design material to the Design Engineer for backcheck and correction. If the Design Engineer does not agree with the checker's notations and the differences cannot readily be resolved between the two, the matter shall be referred to the Group Leader (and Section Head if necessary) for decision. The Design Engineer shall then make all necessary corrections to the design.

5.1.4.3 RECHECK. Upon completion of the backcheck and corrections, the Design Checker shall recheck pertinent portions of the design to determine that all proper corrections have been made. Only when he is satisfied that all corrections have been made and the design is suitable and adequate shall the Design Checker sign the original design calculations.

Jacobs Engineering also provided the following on the levels and other items covered under its current quality assurance/quality control process:

Level 1: Checking Process (applied to calculations, plans, drawings, reports and software input). Typically involves a 100% document check, 100% input check, spot check (or partial check), originate and check, backcheck, update, and recheck.

Level 2: Review Process (applied to concepts, intent, and processes). Typically involves a concept review, spot review, reasonableness review, prepared action plan, and formal peer review report or less formal memorandum.

Level 3: Authorization Process (applied to documents that require signature and/or review by management). Typically involves signature by management on a matrix giving checking and review requirements or on a Job Specific Quality Plan.

Interviews With Former Sverdrup & Parcel Employees. Safety Board investigators interviewed the former Sverdrup & Parcel employee who did the "detailing"[49] of the truss connections of the I-35W bridge (the detailer) as well as the engineer who would have been responsible for checking those designs and design calculations (the checker). Also interviewed was another engineer (other engineer) who worked with the detailer on the bridge design and who may have made some gusset plate calculations.

The detailer said that he joined Sverdrup & Parcel in June 1963, about 2 months after the company had presented the preliminary engineering report for the I-35W bridge to Mn/DOT. He recalled that in November 1963, he was assigned to detail the joints on the truss of the bridge. He said that, at that time, he was an engineer-in-training; he did not receive his professional engineer registration until 1966, after he had left Sverdrup & Parcel.

His main job was to calculate the number of rivets necessary and to develop a rivet pattern, which would govern the in-plane size of the gusset plates. He said that he also did some preliminary sizing of the gusset plate thickness based on the transfer of forces between the chord members. He did not indicate that he had been familiar with the calculations necessary to size the gusset plates for transfer of forces between the chord members and the verticals and diagonals, and said that he assumed this was done by a more senior person after his work was done.

The detailer said that when he needed help, he would ask the checker or the other engineer to assist him. He thought that the person most likely to have done the final gusset plate sizing was the checker because he was the resident expert in detailing structural steel joints. When asked about changes in the steel specifications, the detailer said he was not involved in that and did not remember having to change any calculations based on changes in materials. He recalled that the checker had checked his calculations in or around April 1964 and did not recall having any discussion about the design check.

The other engineer had joined Sverdrup & Parcel in 1957 and remained with the company until his retirement in 1996. He said that he checked some of the design loads on the main span of the I-35W bridge and also checked some of the joints. He also worked on design of the floor truss gusset plates. He said that he and the detailer probably did the calculations on the main truss gusset plates and that these calculations would have been rechecked multiple times by other engineers. Both the other engineer and the checker noted the possibility that several engineers could have participated in doing computations for the I-35W bridge gusset plates. Both also brought up the possibility that the detailer could have been provided with another engineer's work from a different project to use as an example.

The checker worked for Sverdrup & Parcel for 41 years, beginning in April 1951 and ending when he retired in January 1992. He indicated that during his career he had specialized in the detailing and checking of trusses and other steel joints.

---

[49] The detailer is responsible for preparing the details of a truss sufficient for a contractor to prepare shop drawings.

In 1989, he authored a 225-page manual, *Detailing Guide for Structural Steel Joints* (discussed in more detail below), which describes a 14-step process for joint design. The checker said that most of his design work involved through truss (as opposed to deck truss) bridge designs. He said that Sverdrup & Parcel would typically use multiple thinner gusset plates rather than one very thick plate to provide some redundancy and to prevent defects that could occur in thick sections.

The checker commented several times during the interview that gusset plate thickness was governed by shear across the joints, and he indicated that the calculations would have accounted for direct stress and flexure. He also commented that joints are typically stronger than the members they connect.

The checker explained that, in general, the original designer or detailer would provide drawings for him to check. During this process, the calculations for the gusset plates would be checked. If any disagreements were noted between the detailer and checker, the documents and drawings would have been rechecked by the project engineer. If the designs and calculations were correct, they would be provided to the contractor (or, in the case of the I-35W bridge, to Mn/DOT), who would use them to prepare the shop drawings.

Based on an examination of his timesheets from that period, the checker said that he would have performed only preliminary design work on the I-35W bridge and that he did not specifically recall working on the I-35W project. He did say that he believed the detailer was a relatively new employee and that he (the checker) would have been assigned to check the detailer's work. In this instance, the checker would have performed much of the detailer's work to teach him the procedures. The design drawings showing the main truss gusset plates were all drawn by the detailer and checked by the checker.

1989 Sverdrup Corporation Design Manual. Jacobs Engineering, in August 2008, provided the Safety Board with a copy of the 1989 edition of Sverdrup Corporation's *Detailing Guide for Structural Steel Joints*. The foreword to this manual notes that it was preceded by two earlier versions, one published in 1964 and the other in 1969, but no copies of these versions have been located. In general, the manual deals with joints in truss bridges and includes guidelines, examples, and reference materials.

The manual indicates that the detailer was responsible for preparing computations that were to be checked and revised as necessary, with the results used to produce the final design drawings. The manual lays out a 14-step process for the design of joints that carry tension. In general, the steps are the same for joints that carry tension and for joints that carry compression, except that net section properties (accounting for area lost to fastener holes) are used for tension joints, while gross section properties are used for compression joints. The 14 steps are summarized below:

1. Use design loads to calculate the number of fasteners required for verticals and diagonals, based on AASHTO specifications for connections.

2. Calculate the direct force carried across the joint. Because design loads for each member are determined independently as the maximum for that member, the design loads for all members at a joint will generally not be in equilibrium. If the loads are far from equilibrium, an equilibrium case should be calculated that gives the maximum direct force across the joint. If the design loads are not too far out of equilibrium, they can be used directly, making the most conservative choice for each component or combination as necessary.

3. Calculate the stress ranges of the members to determine if fatigue cracking is an issue at the joint.

4. For the chord members, calculate the contribution to the cross-sectional area of the side plates and cover plates to determine the number of fasteners to be used for each type of plate.

5. Calculate initial thicknesses of the splice plates and gusset plates to carry the direct force across the joint. The height of the gusset plate in this case is taken as equal to the height of the chord members. Material used for the splice plates and gusset plates should be the same as for the more highly stressed chord member. The allowable stress is reduced by 20 percent for this calculation.

6. Lay out the joint to calculate the in-plane dimensions of the gusset plates, as determined by the angles of the diagonals, any chamfers, and the approximate fastener patterns. Fastener spacing follows AASHTO specifications.

7. Because the full in-plane sizes of the gusset plates introduce an eccentricity in the joint, check the maximum total stress (direct force across the eccentric joint plus bending stress) on the upper and lower splice plates against the allowable tension or compression stress.

8. Determine the numbers and types of fasteners needed to connect the chord members through the splice plates and gusset plates.

9. Compare the average stress (from the direct force plus bending stress [step 7]) across the gusset plate against the allowable tension or compression stress.

10. Evaluate the fastener patterns against the requirement to maintain net sections.

11. Check to make sure that welded cover plates for diagonals extend far enough inside gusset plate connections to fully develop the stress in the cover plate within the joint.

12. Check the pull-out resistance of tension members through a combination of tension and shear around the end of the member. This check is similar to a block shear calculation.

13. Calculate the average shear stress in the gusset plates along a section between the chord members and the diagonals and vertical, using the design loads or an equilibrium load case that gives the maximum shear stress on that section. Multiply the average shear stress by 1.5 and compare that value to the allowable shear stress, increasing the thickness of the gusset plates if necessary to meet the allowable stress level.

14. Recheck the joint to assess the possibility of fatigue cracking.

In addition to the 14 steps listed above, some additional guidance was also provided, as excerpted and summarized below:

a. Gusset plates should be as small as possible.

b. Use no more than 5–10 percent more fasteners than required.

c. Keep fasteners as symmetrical as possible about the centerline of a member.

d. Use cover plates on diagonals the full length of the member to act as diaphragms between gusset plates. Cutouts in the cover plates might be needed to satisfy this requirement and also allow access to fasteners.

e. Check the unsupported edge distance where there is a possibility of buckling from a compression diagonal member. If the unsupported edge distance exceeds AASHTO specifications, an angle can be added to stiffen the edge.

The manual also includes design guidelines for special types of joints, such as those at the ends of trusses; middle joints between upper and lower chords; and joints for lateral, sway, or portal braces.

Gusset Plate Calculations for Orinoco Bridge. A Jacobs Engineering search of company records found one set of Sverdrup & Parcel documents that showed computations related to bridge main truss gusset plates. The documents were for a Sverdrup & Parcel-designed bridge over the Orinoco River in Venezuela. The Orinoco bridge is a suspension bridge, but it uses a truss to support and stiffen the deck. The excerpted computation sheets that Jacobs Engineering provided to the Safety Board date from May–July 1964, and they show calculations for some of the joints in the main trusses and floor trusses. The excerpts from the design plans that were provided are also dated 1964.

The excerpts included a complete set of computations for joint L9. The numbers of fasteners needed for the vertical and diagonals were first calculated, adhering to AASHTO specifications that connections should carry the average of the design load and the member capacity, but not less than 75 percent of the member capacity. The connections were checked both for tension using net section properties and for compression using gross section properties but a reduced allowable stress to prevent buckling. The direct force plus bending stress in the lower splice plate was checked for both tension and compression; these stresses arise from the direct force across the chord splice coupled with the eccentricity in the joint introduced by the extent of the gusset plates. The direct force across the chord splice was assumed to be the member capacity in tension or compression. The numbers of fasteners needed for the chord members were then calculated. The gusset plate was checked for shear along a section between the chord members and the diagonals and vertical; the maximum average shear stress was compared directly to the allowable stress (if the average shear stress had been multiplied by 1.5, that value would also have been less than the allowable stress). The gusset plate was then checked for bending stress developed along a net section through the top row of rivets on the chord members.

The excerpts also included computation sheets for joint U17, a three-member joint in which the vertical is attached to a continuous upper chord with no diagonal members. These computation sheets included some calculations for connecting lateral braces to the upper chord through a gusset plate. The numbers of rivets required were calculated for each member. The gusset plate for connection of the lateral braces was also checked for shear and bending; in the shear check, the average shear stress was multiplied by 1.5 for comparison with the allowable stress.

Mn/DOT Bridge Design Review Process

Mn/DOT representatives provided the Safety Board with the *Minnesota Highway Department Bridge Design Manual*, dated April 12, 1972, which contained a section documenting the State's process for checking design consultant plans that would have been in place at the time the I-35W bridge was designed. The process involved reviewing and checking for major items to ensure adequate layout control and coordination with roadway plans (for example, control points and strength requirements for railing, slab, beam and girders, bearings, piers and abutments, and pilings). The process did not involve checking the bridge design calculations for improperly designed gusset plates and connections.

Mn/DOT officials said that the current process for checking consultant plans is similar to that of 1972. It still involves reviewing and checking for major items to ensure adequate layout control and coordination with roadway plans. This process is accomplished through a "cursory review" and "thorough check" of bridge design plans. A cursory review generally refers to a comparative analysis to ensure compliance with standard practices and consistency with similar structures. A thorough check generally refers to performing complete mathematical computations to identify discrepancies in the plans. The first cursory review and thorough check are performed at the partial-plan stage to ascertain that the consultant is proceeding in the right direction. The second review and check occur at the final plan stage, when the plans should be sufficiently complete that they can be stamped by the professional engineer. Table 13 lists the items covered in a cursory review and thorough check of both the partial plan and the final plan. None of the items listed in the table involve checking bridge design calculations for improperly designed gusset plates and connections.

Table 13. Current Mn/DOT process for checking consultant bridge plans.

Partial plan	Final plan
Thorough check	**Thorough check**

Horizontal and vertical clearances
Stations and elevations on survey line
Deck and seat elevations at working points
Deck cross-section dimensions
Working line location and data
Coordinates at working points and key stations
Substructure locations by station
Framing plan
Conformance to preliminary plan
Design loads
	Pay items and plan quantities
	Project numbers
	Design data block and rating on grade, profile and
	   estimates sheet
	Job number
	Certification block
	Standard plan notes
	Concrete mix numbers
	Construction joint locations
	Prestressed beam design if inadequate design is
	   suspected
	Bridge seat elevations at working points
	Utilities on bridge
	Existing major utilities near bridge

Cursory review	**Cursory review**

Proposed precast beams
Precast conformance to industry standards
Proposed steel beam sections
	Steel beam splice locations and diaphragm spacing;
	   flange plate thickness increments
	Abutment and pier design, check against consultant's
	   calculations
	Conformance to foundation recommendations
	Pile loads and earth pressures, check against
	   consultant's calculations
	Rebar series increments (min. 3 inches)
	Interior beam seat elevations
	Bottom-of-footing elevations (for adequate cover)
	Railing lengths and metal post spacing (for fit)
	Use of B-details and standard plan sheets
	Conformance to aesthetic requirements
	General, construction, reference notes, etc.
	Quantity items on tabulations
	Precast beam design, check against consultant's
	   calculations

	No check or review required
	Diagonals on layout sheet Figures in bills of reinforcement Bar shapes and dimensions Rebar placement dimensions Bar marks on details against listed bars Quantity values (including total of tabulations)

FHWA Bridge Design Review Process

According to FHWA officials, in the 1960s, FHWA engineers were more likely to be involved in the detailed engineering design of projects and to be active participants during construction. The current FHWA workforce is much smaller than in the 1960s, and the agency has far fewer employees with actual project experience. As a result, current FHWA employees place more attention on broader program delivery activities than on detailed design issues. This approach is consistent with current agency direction, which has been shaped through the years by various transportation laws.

The FHWA and Mn/DOT signed a stewardship plan in December 2007 that sets forth the respective roles and responsibilities of each party in the administration and oversight of the Federal-aid Highway Program in the State of Minnesota. The stewardship plan covered two functional areas: project and program oversight.

The stewardship plan defines project oversight as activities that would be undertaken as part of the project development process, to include the following:

- Environmental process,
- Right-of-way process,
- Design monitoring process,
- Local public agency delegation process,
- Programming and project authorization/agreement processes,
- Intelligent Transportation System process, and
- Construction and contract administration process.

Program oversight is defined as activities that would be undertaken as part of the administration of programs of mutual benefit to Mn/DOT and the FHWA. The activities listed under program oversight include the following:

- Bridge program,
- Financial management,
- Maintenance monitoring,
- Material acceptance,
- Pavement management and design,
- Planning,
- Research, development, and technology,
- Safety and traffic, and
- Miscellaneous programs and activities.

In Minnesota, oversight determinations are made according to the type and cost of projects, as noted below:

- All major bridges on the National Highway System with a cost of more than $10 million have full FHWA oversight.

- Interstate construction or reconstruction projects over $1 million have full FHWA oversight. Highway construction projects on the Interstate System under $1 million are administered by Mn/DOT.

The FHWA Headquarters Bridge Division is responsible for approving preliminary plans for unusual bridges and structures on the Interstate System. Unusual bridges are generally those that have (1) difficult or unique foundation problems, (2) new or complex designs with unique operational or design features, (3) exceptionally long spans, or (4) design that departs from currently recognized acceptable practices. Examples of unusual bridges include cable-stayed, extradose,[50] suspension, arch, segmental concrete, movable, or truss bridges.

FHWA officials provided the Safety Board with a list of the major items to be addressed in a preliminary plan review. These items included:

- Use of high-performance materials,
- Use of new technologies,
- New, innovative materials,
- Opportunities for accelerated construction,
- Unique/creative new uses of known materials,
- Constructability and appropriateness of construction techniques,
- Maintainability,
- Cost-effectiveness,
- Aesthetic requirements,
- Corrosion protection strategy,
- Improved details to eliminate existing problem areas on bridges (such as bridge expansion joints, fatigue-prone details, and bearings),
- Hydraulic/scour analysis and deck drainage,
- Geotechnical requirements, and
- Foundations.

---

[50] An *extradose* bridge combines the structural characteristics of conventional cable bridges and post-tensioned box girder bridges.

The preliminary plan review should also consider the bridge location, length, width, span arrangement and superstructure system including traffic requirements, safety measures, channel configuration, and stream flow. Feasible alternatives for a proposed bridge crossing, along with their merits and shortcomings, should be identified and discussed as well. None of the major items included checking the bridge design calculations for improperly designed gusset plates and connections.

The final plan review is the stage of project development when the plans, specifications, and estimates package is submitted for review and approval. A typical package includes a set of the completely detailed project plan sheets, the project contract proposal, and a copy of the design engineer's construction cost estimate. It may also include other items such as right-of-way certificates or environmental permit applications. The final plan review consists of examining the submitted package for consistency with the project's scope of work, conformity to acceptable engineering design and construction practices, Federal aid eligibility, environmental compliance, and adherence to all appropriate Federal rules and regulations. The review also ensures that all previous comments, such as those made at the preliminary plan review, have been satisfactorily resolved.

Design Review Processes in Other States

In conversations with other private design consultants, Safety Board investigators were told that final design plans were checked internally, with no expectation that additional checks would be necessary once the plans had been delivered to the State. The consultants stated that the review processes employed by State DOTs varied widely and were largely dependent on the preferences and practices of the reviewing personnel.

To evaluate State review processes, Safety Board investigators looked at transportation departments in 14 States. Factors considered when selecting an agency were size, location, and bridge inventory. Investigators interviewed officials and staff of these representative State DOTs to assess their general approaches in designing bridges and in reviewing and approving design plans.

The Safety Board survey found variations among States as to the percentage of bridge design work performed in-house versus outsourcing. Factors that affect such decisions include the number of bridges for which designs are required, the size and makeup of in-house staff, and the complexity of the project. State DOTs that have maintained a strong engineering staff may do almost all design work in-house, while others—because of decentralization or attrition—may outsource all but the most routine design tasks.

When a State contracts with an outside bridge design firm, it retains the responsibility for reviewing the consultant's design and plans at various stages. Typically, the engineering concepts and design plans are examined at 30 percent (preliminary), 60 percent, 90 percent, and 100 percent (final) completion. The State transportation officials interviewed all made a distinction between reviewing plans and checking plans:

- Reviewing plans involves assessing the general suitability of the design and design details and ensuring compliance of the design with the requirements of the project and with good engineering practice. A design review also assesses adherence to budget and schedule as well as compliance with internal procedures.

- Checking plans involves verifying that design assumptions, design computations, drawings, and specifications are complete, correct, and consistent with all job requirements and with good engineering practice.

The officials interviewed for this survey generally agreed that their in-house engineers—working with FHWA engineers if appropriate—review design plans primarily from the standpoint of suitability, constructibility, budgeting, and scheduling, but also consider a design's conformance with sound engineering principles. When designs are checked, the structural elements reviewed do not include gusset plates or connections. For calculations that are not checked, the transportation departments rely on the seal of a State-registered professional engineer. The seal is applied to all approved design plans to certify that all design assumptions, drawings, specifications, calculations, and computations have been checked and are correct.

The use of computer-aided design has changed some quality control functions by making it more difficult for State engineers to check design calculations. For example, Safety Board investigators were told that in the 1960s, when calculations were done manually, State engineers could review each step in the calculations to verify accuracy. Today, however, design calculations are delivered in the form of computer printouts, which show the results of the calculations. Checking the calculations would require that a State engineer confirm that the program inputs (geometry, section properties, and material properties, for example) were correct and were properly entered into the computer. The State review process normally does not extend to this level of verification.

In some cases, State DOTs contract with design consultants to perform an independent review of the designs and plans of other consultants. These independent third-party reviews, which usually include a check of the original design and calculations, are most likely to occur if the project is unusual or complex—for example, if it involves cable-stayed bridges, suspension bridges, truss spans, concrete arch bridges, or bridges requiring unique analytical methods. If the project involves a standard design, such as prestressed concrete beam bridges and deck slab bridges, an independent review is generally not performed.

State DOTs typically use an "error and omissions" contract with consultants to enforce accountability by assessing penalties for poor engineering performance.

If it is determined that an error was caused by the consultant, the consultant must correct the deficiency at no cost to the State as well as assume financial responsibility for any consequences that result from the error.

Safety Board investigators obtained basic information from State DOTs, such as the total number of districts/regions in each State, total number of State bridges and local bridges (including total deck area), total percentage of consultant bridge designs and in-house bridge designs, and types of bridge load rating programs and bridge management systems used. This information is shown in appendix B.

Examples of Bridge Design Errors

As part of the Safety Board's survey of the bridge design approval processes used by States other than Minnesota, investigators inquired as to whether the reviewed States had ever accepted highway bridge designs that were later revealed to contain significant errors. Of the 14 States surveyed, 10 acknowledged having reviewed and approved bridge designs that were later found to be deficient, as summarized below.

A steel girder bridge built in 1977 was discovered 12 years later to have had inadequate reinforcing in some of the pier caps. The State applied post-tensioning to the pier caps found to be in distress. The bridge design firm disputed whether this was a design error, and liability was not established.

In 1997, a load rating analysis performed on a continuous span multigirder bridge that was then under construction revealed that some girder sections had substandard load capacity. A review of the design confirmed an error in design calculations. Cover plates and additional web stiffeners were retrofitted to portions of the girders, and new bearings were installed to provide the required design capacity. The design consultant assumed the cost of the remedial work that was attributable to the design error.

In 2002, the initial load rating of a new continuous welded plate girder bridge revealed that the inventory shear rating was less than the desired design rating because of the spacing of intermediate stiffeners on the girder web. The State resolved the issue by installing additional intermediate web stiffeners in the field. The design consultant was charged for the cost of the new material.

In early 2003, construction began on the first of two bridges (one eastbound and one westbound) that were to be built as two-cell cast-in-place box girders with a center web and two inclined exterior webs. When construction of the first bridge was about 40 percent complete, construction inspectors discovered stress cracks in the webs of the box girders. Additional vertical post-tensioning had to be added to the webs of the remaining segments being cast to prevent additional cracking. A review of the original design by an independent design consultant revealed that the bridge design firm, assuming a simplified distribution of load, had determined that each of the three webs in the box girder would carry one-third of the load. In fact, however, 40 percent of the load was borne by the center web. A retrofit was developed to add external post-tensioning to relieve the stress in the portion of the bridge that had already been built, and the original designer modified the plans for the eastbound bridge to add vertical post-tensioning to the webs. The State DOT had the independent design consultant perform an "over-the-shoulder" review of both the retrofit and the redesign.

Construction began in 2004 on a span-by-span segmental superstructure bridge on single-column piers with single drilled-shaft foundations. According to the State DOT, the design consultant used aggressive drilled-shaft design assumptions despite highly variable conditions in the borings and against the advice of State reviewers. During construction, the drilled-shaft embedment depths were further reduced based on load tests, even though test borings were not performed at the load test sites for comparison with design borings. In the early stages of construction, one pier rapidly settled 11 feet during erection of one of the spans. In an unrelated event, another pier settled 50 percent more than the allowable limit set by the engineer of record. An intensive soil investigation was conducted along with a reevaluation of the drilled-shaft design methods. Each foundation was reviewed, and specific repair details were developed to strengthen or to modify the foundations.

In 2004–2005, some girders of a chorded welded steel plate girder highway bridge were found to have extreme skew, which resulted in differential deflection of the girders during slab pour. The structure also experienced unpredicted thermal movement of the girders due to stiff substructures and complications from failure to account for thrust forces. Expansion bearings were added at certain locations, tilted girders were jacked to near plumb position, thrust blocks were installed at heavily kinked girders to redirect thermal movement, fracture bolts were replaced, and heavier cross frames were installed.

In 2007, some of the girders being used to erect a continuous prestressed concrete beam bridge were found to have been fabricated with fewer stirrups[51] than originally specified because the shop details did not match the plan details. The engineer who checked the shop details did not detect the error until 34 of 66 girders had been fabricated with too few stirrups. The 34 girders as fabricated met the current load and resistance factor design specification, but they did not have the normal reserve capacity. The girders were used, and the fabricator and design consultant were penalized monetarily.

In 2007, a three-lane highway bridge ramp had to be demolished and rebuilt because of a design error. The ramp consisted of cast-in-place concrete retaining walls spaced about 52 feet apart. The retained soil between the walls supported an on-grade cast-in-place post-tensioned cantilevered slab that overhung the walls on both sides by 13 feet 3 inches. The original slab design did not fully account for the soil support, which effectively resulted in a larger overhang. The ramp had to

---

[51] *Stirrups* are steel or carbon-fiber-reinforced polymer bars that are embedded in a concrete member to add shear reinforcement.

be demolished and redesigned to incorporate the soil support condition. Internal as well as external checks were performed on the revised plans. According to the State DOT, this particular portion of the project had been overlooked during the initial review and checking of the design.

In 2008, during construction of a two-span cast-in-place prestressed box girder bridge, State inspectors reviewing shop drawings submitted by the subcontractor discovered that the prestress cable path shown on the drawings was different from that shown on the plans. The design consultant reviewed the design calculations and found that the plans had been in error and that the shop drawings were correct. The State determined that the subcontractor had likely found the error while preparing the shop drawings and had contacted the designer and corrected the mistake before submitting the drawings to the State.

In April 2008, during a roadway widening project that involved several bridges, State inspectors found unexpected cracking in reinforced concrete pier caps to be used for the steel multigirder three-span overpass bridges included in the project. Because of a reinforcing detailing/design error, the primary flexural reinforcement in the top of the hammerhead pier caps had inadequate anchorage and development length. The detailing of the confinement reinforcement did not meet the AASHTO ratio of 0.003 in each direction in each face of the cap. The cap reinforcement also failed to meet AASHTO requirements for temperature and shrinkage. The caps were removed, redesigned, and reconstructed with proper reinforcing detailing.

Mn/DOT Certification of Bridge Inspectors

Federal requirements for bridge inspector qualifications are set forth at 23 CFR 650.309. According to Mn/DOT officials, requirements for certifying State bridge inspectors are based on Federal regulations as well as on State statutes and rules. Mn/DOT may certify an inspector as an assistant bridge inspector or a bridge inspection team leader. Completion of the 1-week training course, "Engineering Concepts for Bridge Inspectors," is required to qualify as an assistant bridge inspector. With this certification, an inspector may only assist in bridge inspections; a certified bridge inspection team leader must be present at the site at all times during the inspection.

A certified bridge inspection team leader can inspect in-service bridges and culverts on State, county, and local highways throughout Minnesota. To qualify as a bridge inspection team leader, a person must meet at least one of the following requirements:
- Be a registered professional engineer in the State of Minnesota,
- Have 5 years of bridge inspection experience,
- Be certified by the National Institute for Certification in Engineering Technologies as a level III or IV bridge safety inspector,

- Have a bachelor's degree in engineering from an accredited college or university, successfully pass the National Council of Examiners for Engineering and Surveying Fundamentals of Engineering examination, and have 2 years of bridge inspection experience, or

- Have an associate's degree in engineering or engineering technology from an accredited college or university and have 4 years of bridge inspection experience.

Additionally, inspectors who meet one or more of the above qualifications must then successfully complete an FHWA-approved comprehensive bridge inspection training course and pass a Mn/DOT field proficiency test before they may become certified as bridge inspection team leaders.

Mn/DOT offers two bridge inspection training courses each year. These courses were developed by the National Highway Institute to meet the definition of a "comprehensive training program in bridge inspection," as defined in the National Bridge Inspection Standards, and are based on the FHWA *Bridge Inspector's Reference Manual.*

The proficiency test for team leaders consists of a routine inspection of an in-service bridge. The inspector is given 2 hours to examine a bridge, take notes, and determine the National Bridge Inventory and Pontis condition ratings. Scoring is based on a scale of 0–100, with a passing score being 70 or more.

Certification of a bridge inspection team leader must be renewed every 4 years. To maintain certification, team leaders must have attended a minimum of two refresher seminars during the preceding 4 years and must have been actively engaged in bridge inspection during at least 2 of the 4 previous years as verified by the supervising engineer.

At the time of the collapse, Mn/DOT employed about 75 bridge inspection team leaders who were responsible for inspecting the 3,500 State highway bridges every 2 years. Of these 75 team leaders, 8 were assigned to regularly perform 200 fracture-critical and special bridge inspections statewide each year. Three were engineers, three were engineering specialists, and two were certified welding inspectors with nondestructive testing certifications.

Minnesota Emergency Preparedness and Lessons Learned

The city of Minneapolis and Hennepin County have emergency operations plans that use the Minnesota Incident Management System to prepare for handling all emergencies within the State. This system was developed based on the National Incident Management System, which is the Nation's first standardized approach to incident management and response, and unifies Federal, State, territorial, tribal, and local lines of government into one coordinated effort. This integrated system establishes uniform response processes, protocols, and procedures for all emergency responders.

According to State emergency management officials, the structure of the system can be established and expanded as necessary for each incident. It allows agencies to communicate using common technology, to share goals and tactical objectives, and to understand the roles and responsibilities of others. The Minneapolis Office of Emergency Preparedness has in place mutual aid agreements with other local municipalities, unincorporated areas, and political subdivisions of the State for reciprocal emergency preparedness aid and assistance in an emergency.

On the day of the bridge collapse, the Minneapolis Emergency Operations Center was opened at 6:20 p.m. to assist in coordinating operations, planning, finance, and logistics for the incident command. Representatives from most agencies involved and city department heads and their ranking officers participated in the activities. The center was staffed 24 hours a day for the first 4 days following the collapse and 12 hours a day until the final victim was recovered on August 20, 2007.

On August 23, 2007, the Minneapolis and Hennepin County Emergency Operations Centers conducted a debriefing/after-action review with all the responding State, city, and county agencies to address issues that arose during the incident response. The meeting identified the following needs:

- Updating the notification system's agency contact list.
- Establishing a mass communication system to assist dispatchers in handling incoming calls.
- Using the city's 311 facility[52] to supplement center resources.
- Establishing an electronic inventory system to track available equipment for emergency use.
- Making provisions for extra cell phone batteries, network printers, electrical and data ports, television monitors, and headphones.
- Establishing an official Web site to provide accurate updates and other incident-related information.
- Establishing better communication within departments to provide updates and directions for employees reporting to work.
- Improving management of assistance to families of victims.

According to the review, the overall incident response was considered a success based on relationships among agencies, open communication, planning, training, and equipment.

---

[52] The 311 system is used for nonemergency public safety and service needs. During the bridge incident, the system relieved 911 operators from handling nonemergency calls. The 311 center had up to 15 operators on duty and eventually was able to screen requests from the public.

ANALYSIS

This analysis begins with a description of the August 1, 2007, collapse of the I-35W bridge, followed by a discussion of the factors that were considered to be potentially causal or contributory to the accident. The analysis concludes by addressing the following safety issues identified during the accident investigation:

- Insufficient bridge design firm quality control procedures for designing bridges, and insufficient Federal and State procedures for reviewing and approving bridge design plans and calculations.

- Lack of guidance for bridge owners with regard to the placement of construction loads on bridges during repair or maintenance activities.

- Exclusion of gusset plates in bridge load rating guidance.

- Lack of inspection guidance for conditions of gusset plate distortion.

- Inadequate use of technologies for accurately assessing the condition of gusset plates on deck truss bridges.

Collapse Sequence

Based on the documented fractures, deformations, and damage patterns on the bridge components, as well as postaccident finite element modeling and the video of a portion of the collapse, the Safety Board developed the likely sequence of events in the collapse of the I-35W bridge. In some cases, the evidence was insufficient to precisely determine the order of events or the cause of specific secondary damage; however, sufficient evidence did exist to establish the major steps in the overall collapse sequence, as described below.

Although the surveillance camera just to the west of the bridge did not capture the beginning of the collapse, it did show that the south end of the bridge center span fractured and began to fall slightly before the north end, indicating that the initial fracture in the two main trusses occurred at the south end of the center span. This fracture area was just out of the camera's view and just south of the U11 and L11 nodes.

Postaccident examination of the bridge components revealed that all four gusset plates at the U10 nodes had fractured into multiple pieces. These fractured gusset plates, coupled with fractures in the lower chord members between the L9 and L10 nodes, completely separated the main trusses in this area, thereby allowing the center span to drop. As part of a determination of the sequence of collapse, the Safety Board evaluated fractures, deformations, damage patterns, and recovery positions of the bridge truss members to understand how the damage was produced and to differentiate between damage that occurred on impact with the ground or the river and damage that occurred before impact.

The evaluations of the main truss members on the south side of the center span showed that all damage patterns and fractures, other than those associated with the gusset plates around the U10 ends of compression diagonals L9/U10, were consistent with secondary damage. For example, the substantial bending deformation associated with the fractures in lower chord members L9/L10 indicated that the L10 ends of these members were being displaced downward before the fractures were created in the L9 ends. The evaluation of the gusset plates at the U10 nodes showed that all four of the plates were fractured in a similar manner, with lateral shifting of the upper ends of the L9/U10 diagonals to the west accompanied by tearing fractures between the compression diagonal L9/U10 and the upper chord, and by tension fractures originating at rivet holes in the area between the L9/U10 diagonal and the U10/L10 vertical member of the node. The gusset plate fractures surrounding the upper ends of the L9/U10 diagonals were then followed by positive bending fractures at the centers of the gusset plates, separating the two members of the upper chord. The evaluation of the fractures in the gusset plates around the U10 end of the compression diagonal showed clear evidence that the node displaced downward, around and through the L9/U10 diagonal, resulting in compression folding of the gusset plate in the area above the diagonal.

The directions of fracture and deformation were consistent with in-line compression loads on the diagonal, and these fractured and damaged areas of the U10 node gusset plates were the only areas in the south side of the center span that met the criteria for an initiating event. The other fracture area in the U10 node gusset plates was between the upper chord members of these nodes; and the features associated with this fracture area were indicative of in-plane bending loads generated as the node dropped, around and through the L9/U10 compression diagonal, subsequent to creation of the other fracture areas in the gusset plates. It should be noted that no evidence was found of corrosion or preexisting cracking on any of the U10 gusset plates.

All of the fractured and damaged areas on the deck truss portion of the bridge were similarly evaluated. Based on the physical evidence of the bridge members, the Safety Board determined that the collapse sequence was as follows:

- The U10 ends of compression diagonals L9/U10 shifted laterally relative to the remainder of the U10 nodes; the gusset plates failed around the ends of the diagonals; and the remainder of the nodes were pulled downward, around and through diagonals L9/U10.

- As the U10 nodes dropped, positive bending loads increased in the portion of the gusset plates that remained attached to the upper chord members of the nodes (U9/U10 and U10/U11). These loads caused the plates to fracture along a vertical line at the center of the nodes.

- With the loss of support from the U10 nodes, floor truss 10 became temporarily suspended from the deck stringers. Large compression loads developed in the upper portion of verticals U10/L10 as the main portion of the U10 nodes moved downward toward the floor truss lower chord attachment on the verticals.

- Tension loads in the lower chord, the lateral bracing, and the deck and stringers pulled the south portion of the deck truss northward and off piers 5 and 6, causing most of the bearing rollers at these piers to fall off the north sides of the piers.

- Downward bending caused lower chord members L9/L10 to fracture adjacent to the L9 nodes.

- At this point, separation of the main trusses in the south fracture area was complete or almost complete, and the south end of the center portion of the truss continued to drop toward the river.

- As the south portion of the truss moved northward toward the river, lower chord members L7/L8 landed on the rollers and the top of pier 6. Lower chord member L7/L8E fractured from the nodes at each end, which allowed the portion of the truss from nodes 8 to nodes 4 to topple toward the east. Lower chord member L7/L8W also landed on the rollers and the top of pier 6, but it did not fracture and remained resting on the top of pier 6.

- On the north side of the center span, the gusset plates around the U10' ends of diagonals L9'/U10' were fractured and deformed in a manner similar to the gusset plates at the U10 nodes, allowing the remaining portions of the U10' nodes to be pulled downward through the diagonals. In addition, compression buckling developed in the lower chords of the main truss between the L11' and L9' nodes.

- As the U10' nodes dropped, positive bending loads increased in the portion of the gusset plates that remained attached to the upper chord members of the nodes (U9'/U10' and U10'/U11'). The loads caused the plates to fracture along a vertical line at the center of these nodes.

- Upper chord members from the U8' to U10' nodes bent downward adjacent to the U8' nodes.

- Lower chord members L9'/L10' fractured adjacent to the L9' nodes from downward bending.

- At this point, separation in the north fracture area was complete or almost complete, and the center portion of the truss dropped into the river, with the south end of the center portion preceding the north end.

- Secondary fractures and damage occurred in the south and north portions of the deck truss outside the center section, and these portions also collapsed, as did the portions of the approach spans that had been supported by the deck truss.

To evaluate loading conditions on the I-35W bridge throughout its life and to identify the associated gusset plate failure mechanisms, the Safety Board enlisted the support of the FHWA, SUNY, and Simulia to conduct a finite element analysis of the deck truss portion of the bridge. This analysis showed that areas of the U10 gusset plates at the end of the L9/U10 diagonals were beyond their yield stress under the dead load of the initial bridge design. As loads on the bridge increased as a result of the added deck (1977) and barriers (1998), the area of the gusset plates beyond the yield stress expanded, but large deflections were prevented by the surrounding elastic material.

With the added construction and traffic loads on the day of the accident, the areas of yielding increased further, and the finite element analysis predicted that the failure mode under these loading conditions would be the unstable lateral shifting of the U10 end of the L9/U10 diagonal. At the point of instability, the lateral shift of the upper end of the L9/U10 diagonal would proceed rapidly, with no increase of load necessary to cause further shifting. The load-carrying capacity of the areas of the gusset plates above the end of the L9/U10 diagonal would be reduced as a result of bending deformation and additional yielding in the gusset plate. Stress would then be expected to transfer to the area between the L9/U10 diagonal and the U10/L10 vertical, resulting in the tensile fractures in the gusset plates emanating from the rivet holes along the lower edge of the L9/U10 diagonal, which were observed in the bridge components. The finite element analysis showed that the lateral shifting instability preceded the tensile fractures emanating from the rivet holes.

The finite element analysis predicted that the lateral shifting instability of the L9/U10 diagonal would have occurred first at the U10W node, which was more highly stressed than the U10E node as a result of the placement of the construction materials. Following the instability and reduction of load-carrying capacity at the U10W node, load would have been shed to the U10E node, triggering a similar pattern of deformation and fractures, after which the failure likely proceeded rapidly, and more or less simultaneously, through both the U10E and U10W nodes.

The inclusion of bowed gusset plates in the models consistent with the preaccident photographs of the U10 gusset plates had two important consequences. First, the load required to trigger the unstable shift of the upper end of the L9/U10W diagonal decreased when compared to models with unbowed gusset plates. Second, when the appropriate bowing was included in the model, the upper end of the L9/U10W diagonal shifted to the west, matching the evidence from the wreckage. With unbowed gusset plates, the L9/U10W diagonal shifted to the east. In summary, with the inclusion of the bowed gusset plates, the finite element model predictions of the failure mode are consistent with the physical observations of the fractures and damage patterns found on the U10 nodes. The bowed gusset plates also decreased the load-carrying capacity of the structure.

The Safety Board therefore concludes that the initiating event in the collapse of the I-35W bridge was a lateral shifting instability of the upper end of the

L9/U10W diagonal member and the subsequent failure of the U10 node gusset plates on the center portion of the deck truss.

The deck truss structure of the I-35W bridge was non-load-path-redundant, which means that it would lose its entire load-carrying capacity if a single primary load member failed. In this accident, the failure of the U10 gusset plates led to the sequential separation of the structural members that had been connected through the plates, which placed unsupportable loads on the remainder of the structure. The Safety Board concludes that, because the deck truss portion of the I-35W bridge was non-load-path-redundant, the total collapse of the deck truss was likely once the gusset plates at the U10 nodes failed.

Other Possible Collapse Scenarios Considered

The Safety Board considered a number of potential explanations for the collapse of the I-35W bridge before determining, as discussed in the previous section, that the collapse initiated with the failure of the gusset plates at the U10 nodes. This section describes alternative causes that were considered.

Corrosion Damage in Gusset Plates at L11 Nodes

Corrosion damage was present on all four of the gusset plates at the L11 nodes and was concentrated along a line at the top of the upper surface of the main truss lower chord. The measured section loss from the corrosion was 5–10 percent for the outside gusset plates and 15–20 percent for the inside (toward the center of the bridge) gusset plates. The Safety Board considered whether this damage could have caused a failure in the L11 gusset plates. As discussed below, none of the available evidence (video recording, evaluation of the fracture and deformation patterns, and finite element analysis) indicated a possible initial failure associated with the L11 nodes. Additionally, an initial failure at the L11 nodes would have produced fractures and deformation in the area north of the U10 nodes, rather than the observed fractures and deformation south of those nodes. All fractures and deformation in the L11 nodes were consistent with impact with the riverbed. These factors are discussed in more detail below.

Video Evidence. The video recording of the collapse shows that the L11W node remained intact well after multiple fractures had occurred in the south fracture area, indicating that the collapse sequence could not have initiated at this node. Although the L11E node was not visible in the video recording, an initial failure at this node—while node L11W remained intact throughout the collapse— would have been expected to produce significant side-to-side tilting or rolling of the bridge deck as the structure fell. The video recording showed that the center portion of the deck truss fell into the river in a generally level orientation, without observable tilting east to west.

Fracture and Deformation Patterns. The Safety Board considered three potential failure modes involving the L11 gusset plates: (1) the gusset plates around the L11 end of tension diagonal U10/L11 could fracture under the tension loading, pulling the diagonal out of the node; (2) the gusset plates around the L11 end of compression diagonal L11/U12 could fail, pushing the diagonal into the node; or (3) the compression diagonal L11/U12 and vertical member U11/L11 together could be pushed into the node, with gusset plate failure around the ends of these members.

Fracture patterns in the gusset plates around diagonals U10/L11 indicated that neither of these tension diagonals pulled from the nodes in tension. The gusset plate fractures around these diagonals contained evidence of bending, and the lower ends of these members appeared to slightly penetrate the nodes and then fold on top of the lower chord members L10/L11 after translating to the side. In addition, the net section loss from corrosion was less than the net section loss from the top row of rivet holes along the lower chord member L10/L11, just below the U10/L11 tension diagonal; the nominal 1-inch-diameter rivets were spaced at 3.75 inches, for a net section loss of 27 percent.

The gusset plates around the L11 ends of compression diagonals L11/U12 contained significant deformation, fracture, and damage consistent with these members having penetrated through or around the L11 nodes. If this damage had been the initial failure, the main truss segment between nodes 11 and 12 would have been unstable, with the U12 and L12 nodes dropping relative to the U11 and L11 nodes. This relative motion would have introduced large in-plane negative bending loads into the portions of the L11 gusset plates attached to lower chord members L10/L11 and L11/L12. But these gusset plates did not fracture in the area of the lower chords, indicating that no bending loads were associated with initial failure of the gusset plates around the L11 end of the compression diagonals.

If the initial failure had been compression diagonal L11/U12 and vertical member U11/L11 together being pushed into the node, the main truss segment bounded by the U10, U12, L12, and L11 nodes would have been unstable, with the U12 and L12 nodes dropping relative to the U10 and L11 nodes. This relative motion would have introduced large in-plane negative bending loads into the portions of the gusset plates attached to the lower chord members L10/L11 and L11/L12, similar to the bending loads that would have been generated if only the compression diagonal had penetrated the node. However, these gusset plates did not fracture in the area of the lower chords, indicating that no bending loads were associated with initial failure of the gusset plates around the L11 end of the compression diagonal and vertical. Furthermore, the outside (east) gusset plate at the L11E node was not fractured in the area between the vertical member and the tension diagonal, confirming that this failure mode did not occur at the L11E node.

Finally, any of these three scenarios would have led to unloading of the L9/U10 diagonals, making unlikely the lateral shifting instability and fractures that were observed in the U10 gusset plates around the ends of the L9/U10 compression diagonals. These three scenarios would also have led to downward motion of the structure north of U10 relative to U10, which would have created negative bending loads in the upper chords through the U10 nodes. In fact, however, the portions of the U10 gusset plates attached to the upper chord members failed under positive, not negative, bending loads.

Finite Element Analysis. Investigators conducted a detailed finite element analysis of the L11 nodes, incorporating areas of reduced gusset plate thickness to represent the corrosion that was found on the L11 gusset plates. The thickness of each inside and outside gusset plate was locally reduced by 0.1 inch, corresponding to a net section loss of 20 percent, consistent with the maximum measured net section loss. The analysis showed that the maximum stress in the gusset plates of the L11 nodes with corrosion was still less than the maximum stress in the U10W gusset plates. The finite element analysis predicted that the L11 node gusset plates, even in their corroded condition, would have been capable of supporting much higher loads than the loads that initiated failure at the U10W node.

Fracture of Floor Truss

Safety Board investigators considered the possibility that the collapse of the deck truss could have resulted from an initial pure tension or pure compression failure of a member within a floor truss. However, examination of the floor trusses reconstructed at Bohemian Flats revealed none of these types of fractures. Although a portion of a fracture in the upper chord of floor truss 10 contained brittle fracture characteristics, this fracture was consistent with bending loads applied when diagonal L9/U10E struck the lower surface of the upper chord of the floor truss, clearly a secondary event.

Preexisting Cracking

Over a period of years before the collapse, bridge inspectors had found cracks of various sizes in the bridge superstructure. Although most of these cracks were in the approach spans, two cracks had been found in the deck truss portion at the site of welds. Neither of these cracks, however, was in an area associated with the identified area of the collapse initiation, and no evidence was found that preexisting cracking contributed to the collapse or significantly affected the collapse sequence.

All fractures in the portions of the deck truss laid out at Bohemian Flats were examined in detail for areas of fatigue cracking, but none were found. In particular, the fractures in the gusset plates at the U10 nodes were typical of ductile overstress tension, shear, and bending consistent with loading on the plates. These gusset plates showed no evidence of fatigue cracking.

Temperature Effects

The deck truss portion of the bridge was designed and constructed with a fixed bearing at pier 7 (on the north side of the river) and roller bearings at the remaining three piers (5, 6, and 8). Postcollapse examination of the roller bearing components showed the presence of roller wear marks, indicating that the deck truss was moving relative to piers 5, 6, and 8 in response to thermal contraction and expansion. This movement limited the amount of longitudinal force applied to the top of any pier.

On the day of the bridge collapse, the temperature had increased approximately 20° F from morning to early evening, the time of the accident. The finite element analysis incorporated this temperature increase with fixed bearings (but allowing for pier flexibility) to evaluate worst case temperature effects. The effects of a difference in temperature on the east and west trusses arising from the position of the sun were also evaluated. This analysis showed that the loads necessary to initiate the lateral shifting instability of diagonal L9/U10W increased with increasing temperature, and this result did not change when differential temperature was included, indicating that the change in temperature on the day of the accident did not play a role in initiation of the collapse.

Pier Movement

Postcollapse survey measurements indicated that piers 5 and 6 exhibited no settlement or displacement, but that piers 7 and 8 were tilting about 9° southward toward the river. The Safety Board evaluated the possibility that movement of one or both of these tilted piers initiated or affected the collapse. The deck truss portion was supported by roller bearings at piers 5, 6, and 8 and by a fixed bearing at pier 7. The lack of movement of piers 5 and 6—coupled with the location of the roller wear marks approximately in the center of the contact plates at piers 5, 6, and 8—established that piers 7 and 8 had no significant longitudinal movement relative to piers 5 and 6 before the accident.

Postcollapse evaluation of damaged piers 7 and 8 showed that their tilted positions occurred because of separations above the bases of the piers. Pier 7 hinged about the top of the pier footing, and the pier 8 columns hinged about a section approximately 3.5 feet above the top of the footings. There was no evidence that the bases of these piers shifted. Thus, the tilting of piers 7 and 8 was a secondary event.

The Safety Board concludes that the examination of the collapsed structure, the finite element analysis, and the video recording of the collapse showed that the following were neither causal nor contributory to the collapse of the I-35W bridge: corrosion damage found on the gusset plates at the L11 nodes and elsewhere, fracture of a floor truss, preexisting cracking in the bridge deck truss or approach spans, temperature effects, or shifting of the piers.

Emergency Response

Minnesota State Patrol dispatchers were notified of the accident by a cellular caller through the 911 system at 6:05 p.m. State patrol dispatchers contacted Minneapolis dispatch, which sent out the first distress call at 6:07 p.m., requesting that all available emergency assistance respond to the I-35W bridge. Within 4 minutes of the call from dispatch, the first Minneapolis Police Department squad arrived on scene, and the first of 19 engine units from the Minneapolis Fire Department arrived. The Hennepin County Sheriff's Office river rescue squad arrived on the river adjacent to the scene at 6:14 p.m. to begin search and rescue efforts.

Minneapolis uses the Unified Command System for responding to emergencies, with the type of response required determining who will serve as the incident commander. In this incident, the assistant fire chief of the Minneapolis Fire Department was the incident commander and his was the lead agency responsible for overall operations and for issues related to the bridge itself. The Minneapolis Police Department was responsible for the investigation on land and for scene security. The Hennepin County Sheriff's Office was responsible for river rescue and recovery. The Hennepin County Medical Center ambulance service was in charge of emergency medical service operations.

About 25 hours after the collapse, the area became classified as a crime scene, and the Minneapolis Fire Department handed over incident command to the Minneapolis Police Department. The Hennepin County Sheriff's river rescue squad continued to be responsible for locating submerged vehicles with the help of the FBI and Navy underwater search and evidence response divers. The Safety Board concludes that the initial emergency response to the bridge collapse by fire and rescue units was timely and appropriate, and the incident command system was well coordinated. The Safety Board further concludes that the damage to bridge components that occurred during victim recovery did not, in this case, prevent determination of the collapse sequence.

Design of the Main Truss Gusset Plates

Based on early indications of the possibility that the I-35W bridge collapse initiated with a failure of the gusset plates at the U10 nodes, the FHWA evaluated the stresses that would have been imposed on the gusset plates by the member design loads (demand) and compared those stresses to the AASHO-specified allowable stresses for the gusset plate materials (capacity). The evaluations were done using methodology consistent with that used by Sverdrup & Parcel for the floor truss gusset plates of the I-35W bridge and consistent with other truss designs. Two critical gusset plate sections were considered—one a horizontal section at the lower edge of the upper chord (or upper edge of the lower chord), and one a vertical section adjacent to the vertical member of the node. Using this approach, a demand-to-capacity ratio (D/C) of 1 indicates that the amount of reserve capacity called out in the AASHO guidelines is exactly met in the design, less than 1 indicates that more reserve capacity is designed than is required, and greater than 1 indicates that less reserve capacity is designed than is required.

The FHWA evaluation found that multiple gusset plates in the deck truss had D/C ratios greater than 1 for three types of loading—shear, principal tension, and principal compression on both the horizontal and vertical sections. Although engineering judgment may allow a structure to be designed with components having a D/C ratio slightly greater than 1, D/C ratios on the order seen for the U4, U10, and L11 gusset plates (all with D/C ratios over 2 for shear) clearly indicate inadequate design capacity. Based on the FHWA evaluation report, 24 gusset plates at the U4, U4', U10, U10', L11, and L11' nodes should have been approximately 1 inch thick—twice their specified 0.5 inch thickness—to have acceptable D/C ratios. The FHWA evaluation also found that one of the edges of the U10 gusset plates (the edge between the L9/U10 compression diagonal and the upper chord) should have been stiffened to be in compliance with AASHO guidelines. Multiple other main truss gusset plates were also found to have D/C ratios greater than 1, indicating that they also had insufficient load capacity.

As previously discussed, the loading and stress conditions of the U10 and L11 node gusset plates in the as-designed bridge were also evaluated using finite element analysis. This analysis showed that portions of these gusset plates were beyond the yield stress of the material under the dead load of the original bridge design, even before any modifications increased the weight of the bridge. When the weight was increased in 1977 and again in 1998 as a result of added deck thickness and modified barriers, the areas of yielding in the gusset plates expanded.

The intent of the AASHO specifications used for design of the bridge would have been to limit the stress in the U10 gusset plates to less than 55 percent of their yield stress under the original bridge design load—which includes dead load, live load, and impact load. Therefore, under only the dead load, which is a portion of the total load, the stress in a properly designed U10 gusset plate should have been substantially less than 55 percent of its yield stress. Thus, the finite element analysis finding that areas of the gusset plate were beyond the yield stress under only the dead load component of the original bridge design is definite confirmation that the gusset plates at U10 had inadequate capacity. The Safety Board concludes that the gusset plates at the U10 nodes, where the collapse initiated, had inadequate capacity for the expected loads on the structure, even in the original as-designed condition.

Origin of the Inadequate Gusset Plates

The Safety Board considered whether the inadequate capacity of the U10 gusset plates was due to errors in design, errors in fabrication, or errors in construction.

A comparison of the physical bridge components with the approved design plans, the component shop drawings, and the construction scheme for the bridge revealed that the main truss gusset plates had been fabricated and installed with no apparent deviations. The Safety Board concludes that, because the bridge's main truss gusset plates had been fabricated and installed as the designers specified, the inadequate capacity of the U10 node gusset plates had to have been the result of an error on the part of the bridge design firm.

The Safety Board reviewed the available design documentation for the bridge in an effort to determine how the bridge design firm arrived at the specifications for the deficient gusset plates. The Board explored the possibility that a materials substitution was improperly implemented, that the calculations for the gusset plates were done incorrectly and inadequately checked, or that some or all of the gusset plate calculations were not done and that this omission was not corrected in the design review process.

Materials Substitution

Computation sheets and design documents obtained by the Safety Board indicated that the preliminary design of the bridge called for the use of high-strength T-1 steel in about half of the truss members and many of the main truss gusset plates. When Mn/DOT, the FHWA, and the designer agreed that T-1 steel would not be used in truss members, the bridge design firm redesigned the components for which it had originally intended to use T-1 steel because these components would have had to be thicker to meet the reduced allowable stress requirements of the lower strength steel. The Safety Board found no evidence that the gusset plates at the U10(') and L11(') nodes had ever been intended to be fabricated from T-1 steel or that their specifications (0.5-inch-thick A441 steel) had changed from the earliest design documents through fabrication and installation. The inadequate capacity of the gusset plates was determined not to have resulted from a failure of the design firm to redesign the plates because of a change in materials.

Design Calculations

As part of the contract with Mn/DOT, Sverdrup & Parcel was required to submit checked design calculations for all aspects of the bridge. Both Jacobs Engineering (successor to Sverdrup & Parcel) and Mn/DOT were able to provide the Safety Board with the checked computations for many aspects of the bridge; however, neither organization could produce any original checked design calculations for the main truss gusset plates. The checked calculations for the floor truss gusset plates were available and indicated that the designer used appropriate design methodologies. The then-current AASHO *Standard Specifications for Highway Bridges*, 1961 edition, required consideration of "shear, direct stress, and flexure," and the floor truss gusset design methodology used by Sverdrup & Parcel included calculations for shear stress as required.

The only reference to design of the main truss gusset plates in the available documentation was in unchecked computation sheets for the preliminary design, which contained calculations to determine the number and spacing of the rivets in the gusset plates. The computation sheets also included calculations of stresses developed in the gusset plates by the transfer of forces between chord members. In some cases, the initial design of the gusset plates was verified as acceptable; and in other cases, the design was altered, with changes to the thickness or number of gusset plates indicated to reduce the stress level. The calculations for gusset plate thickness used only the forces that were expected to pass across the splices between chord members, without regard to the shearing forces introduced to the gusset plates by the diagonal and vertical members, as required by the AASHO specifications. Using this methodology alone, the gusset plates at the U10(') and L11(') nodes (which did not change from the preliminary to the final design) were sized conservatively. However, at the U10(') and L11(') nodes, the shearing forces were much larger than the chord splice tensile forces, and the chord splice load methodology was inadequate by itself to produce an appropriately sized gusset plate. Particularly at the U4('), U10('), and L11(') nodes, additional shear stress calculations such as those performed for the floor truss gusset plates would have been necessary to ensure that those main truss gusset plates were properly sized.

Although design details of various gusset plates were changed from the time the computation sheets for the preliminary design were prepared until the final design, the thickness and material of the U4('), U10('), and L11(') node gusset plates did not change—which resulted in the gusset plates at these three nodes being substantially underdesigned. An evaluation of the final design showed that the gusset plates at multiple other nodes also were underdesigned, though not to the extent of those at the U4('), U10('), and L11(') nodes. The fact that the U4('), U10('), and L11(') nodes had gusset plates with a basic and very serious design error—and that additional nodes had gusset plates with inappropriate thicknesses—suggests that the design error was not the result of a single calculation error associated with a specific node. Furthermore, the shear calculations that were done for the floor trusses of the I-35W bridge and the main truss calculations done for the Orinoco bridge demonstrated that Sverdrup & Parcel knew how to properly apply these calculations. These facts, coupled with the lack of any documentation for main truss gusset plate calculations, indicate that none of these plates were designed correctly because the appropriate calculations were simply not made for these design elements. Therefore, the Safety Board concludes that, even though the bridge design firm knew how to correctly calculate the effects of stress in gusset plates, it failed to perform all necessary calculations for the main truss gusset plates of the I-35W bridge, resulting in some of the gusset plates having inadequate capacity, most significantly at the U4('), U10('), and L11(') nodes.

The AASHO specifications also require that unsupported edges of members be stiffened if the ratio of the unsupported length to the thickness of the member exceeds 48. For the U10 gusset plates, the edge between diagonal L9/U10 and the upper chord had an unsupported length of 30 inches and a thickness of 0.5 inch, meaning that the ratio between these two values was 60, exceeding the ratio above which stiffening was required. The failure of the design to include stiffening at this location is further evidence of improper design of the main truss gusset plates. However, even if the specification had been followed, it would have had little effect on the D/C ratio or the adequacy of the gusset plates; and the portions of the U10 gusset plates that ultimately failed, which surrounded the ends of the compression diagonals, would have continued to remain above the yield stress of the material. For this reason, the addition of edge stiffeners would not have made the U10 gusset plates adequate or prevented them from yielding. The Safety Board concludes that although the U10 gusset plates would have required edge stiffeners according to AASHO specifications, the addition of stiffeners would not have made the U10 gusset plates adequate or prevented the gusset plates from yielding.

Design Quality Control

Bridge design firms should have appropriate quality control procedures in place to ensure that design errors arising from any source are identified and corrected. The quality control procedures should ensure that appropriate calculations are made and that these calculations receive the appropriate review and check by qualified individuals. Jacobs Engineering provided the Safety Board with the design review procedures that would likely have been similar to those used by Sverdrup & Parcel at the time the bridge was designed. This process involved several iterations of design checking, backchecking, and rechecking until a final design document was produced. Jacobs Engineering also provided the Safety Board with its current (since April 1975) quality control and design checking procedures, Sverdrup & Parcel's *Quality Control Coordination and Checking Procedures*. Although these procedures are more detailed than the earlier ones in terms of the specific individuals who would participate in the review, the overall procedures themselves appear to have changed little. The procedures appear to be sufficient to ensure that calculations receive appropriate levels of review; they also specify the types of calculations necessary for gusset plates but contain no explicit procedure for ensuring that all necessary calculations are performed.

If the appropriate calculations had been generated for the main truss gusset plates, there is no reason to believe that they would not have been subjected to the same checking and quality control process that would have applied to any other part of the structure. The Safety Board concludes that the design review process used by the bridge design firm was inadequate in that it did not detect and correct the error in design of the gusset plates at the U4('), U10('), and L11(') nodes before the plans were made final.

FHWA, Mn/DOT, and Other State DOTs

Mn/DOT and the FHWA (actually, their predecessor organizations) were closely involved in some aspects of the design process for the I-35W bridge, which began with the bridge design firm's initial designs in 1962 and ended with acceptance of the final design in 1965.

In March 1964, based on Mn/DOT and FHWA concerns, the bridge design firm eliminated T-1 steel from all structural members. This change required a redesign of all the members originally specified as T-1 steel.

The FHWA also took issue with the design of a typical node. The agency was concerned about the lack of symmetry in the rivet patterns and about the configuration of the ends of some of the structural members at the node. The bridge design firm changed the design to address these concerns.

Although both Mn/DOT and the FHWA were closely involved with some specific features of the I-35W bridge design, neither organization detected the failure to perform the appropriate design calculations for the gusset plates in the main trusses. At that time, Mn/DOT had design review procedures that provided for checking of consultants' computations—but these procedures were not applied to gusset plates. Complex construction projects are often contracted out to design firms, such as Sverdrup & Parcel, because neither the State nor the FHWA has sufficient resources. This same lack of resources causes State and Federal authorities to rely on the stamp of a professional engineer to certify that all design computations are appropriate, complete, and accurate.

The Safety Board concludes that neither Federal nor State authorities evaluated the design of the gusset plates for the I-35W bridge in sufficient detail during the design and acceptance process to detect the design errors in the plates, nor was it standard practice for them to do so.

This investigation revealed a number of other instances, not involving gusset plates, in which questionable bridge designs have been certified by a designer and reviewed and approved at both the State and Federal levels. For example, of the 14 State DOTs surveyed by the Safety Board with regard to their design review and approval processes, 10 acknowledged having approved bridge designs that were later found to be deficient. All but one of these deficient designs had been approved within the past 10 years, most within the past 6 years. The design errors ranged from girder sections that had substandard capacity due to errors in design calculations, to incorrect loading assumptions in the design of box girders, to inaccurate shop drawings. In most cases, the errors were revealed during initial construction; however, in one case, a deficiency was not discovered until the bridge had been in service for several years. In each case, redesigns or retrofits were required to address the errors. The Safety Board concludes that current Federal and State design review procedures are inadequate to detect design errors in bridges.

The number of deficient bridge designs identified through the Safety Board's relatively limited sampling suggests that design errors in bridges are not restricted to a particular bridge type, a particular State, or a particular time frame, and instead reflect a deficiency in the processes used for reviewing and approving the designs. The Safety Board believes that the FHWA should develop and implement, in conjunction with AASHTO, a bridge design quality assurance/quality control program, to be used by the States and other bridge owners, that includes procedures to detect and correct bridge design errors before the design plans are made final; and, at a minimum, provides a means for verifying that the appropriate design calculations have been performed, that the calculations are accurate, and that the specifications for the load-carrying members are adequate with regard to the expected service loads of the structure. The Safety Board notes that, as a result of this accident, Mn/DOT has revised its *LRFD Bridge Design Manual* to require an independent review of all major bridges designed by consultants.

Bridge Load Rating

Bridge load ratings are most commonly performed when relevant changes in condition are observed or when any new weight (dead load) is added to the structure. Initially, the design firm provided Mn/DOT with the capacity for each member in the truss and for both approach spans (however, no data were found regarding the capacity of the gusset plates); this information was most likely used as the basis for the 1979 load rating that was performed in conjunction with the 1977 construction project, which had added dead weight to the bridge. Mn/DOT performed another load rating that coincided with the 1998 modifications to the bridge's median barrier and outside traffic railings. Both of these load ratings were believed to have been conducted in accordance with the National Bridge Inspection Standards and adhered to the requirements for following AASHTO guidance. However, because the guidance did not consider the connections (gusset plates) and did not provide information on how to evaluate gusset plates, the gusset plates were never evaluated over the life of the bridge. In 2007, Mn/DOT instituted a policy under which a load rating must be performed on any new bridge before it is opened to traffic. This requirement is in contrast to AASHTO guidance, which directs bridge owners to load rate their bridges only when a significant change occurs.

Following its opening in 1967, the I-35W bridge became subject to the AASHTO load rating guidance in conjunction with the newly established National Bridge Inspection Standards, which had been put forth in 1971. As a result, Mn/DOT performed the first load rating of the I-35W bridge in 1979, 12 years after the bridge was put into service. Had a load rating been performed before the bridge was opened, and had it included an evaluation of the connections (gusset plates), the design error might have been detected, and this accident would not have occurred. The Safety Board concludes that, because current AASHTO guidance directs bridge owners to rate their bridges when significant changes occur but not before they place new bridges in service, the load-carrying capacity of new bridges may not be verified before they are opened to traffic. The Safety Board believes that AASHTO should revise its *Manual for Bridge Evaluation* to include guidance for conducting load ratings on new bridges before they are placed in service.

Had an evaluation of gusset plate capacity been included in any of these load ratings, the analyses should have revealed the improperly designed gusset plates, but such ratings did not consider the strength of the connections. The Safety Board concludes that, had AASHTO guidance included gusset plates in load ratings, there would have been multiple opportunities to detect the inadequate capacity of the U10 gusset plates of the I-35W bridge deck truss.

The fact that gusset plates were not considered in load ratings is probably reflective of the tendency to assume that they are stronger than the members they connect, an assumption supported by the apparent lack of previous bridge collapses involving gusset plates. This tendency also probably explains why two of the more commonly employed bridge load rating software programs—BARS and Virtis—do not incorporate the strength of connections in their analyses. The exclusion of variables not considered important to load rating, such as gusset plates, allows programmers and engineers to simplify the analysis without seemingly affecting its accuracy.

As part of the survey of 14 representative State DOTs, Safety Board investigators obtained information on the types of bridge load rating computer programs currently in use. (See appendix B.) The survey revealed that as many as 15 bridge rating programs, in addition to BARS and Virtis, are in common use. At the time of the collapse, none of these computer programs considered the strength of connections (gusset plates). The assumption that gusset plates are stronger than their members appears to prevail despite the fact that, as indicated in the problem statement to the proposed FHWA–AASHTO joint study of gusset plates, the complex geometry and stress in those connections present bridge engineers with special analytical and design challenges that have not previously been adequately addressed.

As stated above, the Safety Board is not aware of a previous highway bridge collapse that resulted from improperly designed gusset plates; however, the I-35W bridge had been in service for 40 years before the deficiency became known. If gusset plates are left out of a load rating analysis, bridge owners do not have the opportunity to verify the original design of these critical components or to account for deterioration of the gusset plates, such as might have occurred through corrosion. Had gusset plates been included in the 1979 and 1997 load rating analyses of the I-35W bridge, Mn/DOT might have determined that the gusset plates at U10 and L11 were in fact the weakest points of the bridge. Instead, Mn/DOT believed that the weakest point of the bridge was in the south approach span and not on the truss portion of the bridge. The Safety Board therefore concludes that because bridge owners generally consider gusset plates to be designed more conservatively than the other members of a truss, because AASHTO provides no specific guidance for the inspection of gusset plates, and because commonly used computer programs for load rating analysis do not include gusset plates, bridge owners typically ignore gusset plates when performing load ratings, and the resulting load ratings might not accurately reflect the actual capacity of the structure.

On January 15, 2008, the Safety Board[53] issued the following safety recommendation to the FHWA:

H-08-1

For all non-load-path-redundant steel truss bridges within the National Bridge Inventory, require that bridge owners conduct load capacity calculations to verify that the stress levels in all structural elements, including gusset plates, remain within applicable requirements whenever planned modifications or operational changes may significantly increase stresses.

Safety Recommendation H-08-1 is currently classified "Open—Acceptable Response."

Also on January 15, 2008, the FHWA issued Technical Advisory T 5140.29, "Load-carrying Capacity Considerations of Gusset Plates in Non-load-path-redundant Steel Truss Bridges," which referenced Safety Recommendation H-08-1 and advised bridge owners to take certain actions to supplement the AASHTO *Manual for Condition Evaluation of Bridges*. For new or replaced non-load-path-redundant steel truss bridges, bridge owners were "strongly encouraged to check the capacity of gusset plates as part of the initial load ratings." For existing non-load-path-redundant steel truss bridges, bridge owners were "strongly encouraged to check the capacity of gusset plates" when performing load ratings as a result of changes in bridge condition or dead load, before making permit or posting decisions, or when necessary to account for bridge alterations that would increase stress levels in the structure. Finally, bridge owners were advised to review previous load rating calculations to ensure that the capacities of gusset plates had been adequately considered.

In May 2008, the FHWA and AASHTO proposed a joint study of gusset plates, with the intent, among other things, of further developing and refining the guidance for bridge engineers in the proper design and rating of gusset plates, and of developing "guidelines, specifications, and examples for the load and resistance factor design and rating of gusset connections."

The Safety Board finds both of these timely responses commendable and takes particular note of the efforts of both the FHWA and AASHTO in providing technical assistance and guidance to FHWA field offices, bridge owners, and State

---

[53] As discussed in appendix C, Safety Board accident investigations and the resulting safety recommendations have played a prominent role in the development of Federal bridge inspection requirements and procedures.

DOTs in the load rating and evaluation of gusset plates of steel truss bridges. But while acknowledging the short-term effectiveness of the FHWA technical advisory, the Safety Board is concerned about the long-term implementation of the second action item in the advisory:

> (2) **Future recalculations of load capacity on existing non-load-path-redundant steel truss bridges.** Bridge owners are strongly encouraged to check the capacity of gusset plates as part of the load rating calculations conducted to reflect changes in condition or dead load, to make permit or posting decisions, or to account for structural modifications or other alterations that result in significant changes in stress levels.

In the view of the Safety Board, this guidance would go further in preventing another gusset-plate-related catastrophic bridge collapse if it were codified through rulemaking or through appropriate guidance documents. Because the National Bridge Inspection Standards incorporate by reference[54] the AASHTO *Manual for Condition Evaluation of Bridges,* in 23 CFR 650.313(c), a provision in that manual would have, for State bridge authorities, the force of a regulation. However, though the *Manual for Condition Evaluation of Bridges* was current at the time of the bridge collapse, it has since been replaced by the recently adopted *Manual for Bridge Evaluation.* The Safety Board therefore believes that AASHTO should modify the guidance and procedures in its *Manual for Bridge Evaluation* to include evaluating the capacity of gusset plates as part of the load rating calculations performed for non-load-path-redundant steel truss bridges. The Safety Board further believes that, when the findings of the FHWA–AASHTO joint study on gusset plates become available, AASHTO should update the *Manual for Bridge Evaluation* accordingly.

Effect of Added Loads Over Time

At the time of the collapse, construction materials, equipment, and vehicles were concentrated along the south portion of the bridge center span above the U10 nodes, where the collapse initiated. The Safety Board evaluated the effect of the concentrated construction loads in conjunction with the effects of the dead load of the original bridge design and the increases in dead load from modifications in 1977 and 1998. The FHWA global finite element model was used to evaluate member loads under a series of load steps tracking the history of changes to the bridge over time.

The collapse of the bridge was initiated when the highly stressed U10W gusset plates were unable to prevent the unstable lateral shift of the upper end of the L9/U10W diagonal, which was driven by the large compressive load (more than 2 million pounds) in that diagonal. The FHWA global model showed that

---

[54] *Incorporation by reference* is a technique Federal agencies use to include and make enforceable material published elsewhere without republishing those materials in full within the regulations. This technique is typically used to incorporate widely used industry-developed codes and guidance.

the dead load of the original bridge design contributed about 73 percent of the calculated load in the L9/U10W diagonal at the time of the collapse. The 1977 increase in deck thickness contributed another 13 percent of that load, and the 1998 modification to the barriers contributed about 5 percent. The milled-off deck in the southbound lanes reduced the load by 3 percent, but this was partially offset by the 2 percent of the load contributed by traffic at the time of the collapse. The construction materials, equipment, and vehicles contributed about 11 percent of the load in the L9/U10W diagonal. Therefore, though the construction materials, equipment, and vehicles added only about 3 percent to the weight of the bridge, they contributed about 11 percent to the load in the L9/U10W diagonal because of their concentrated location.

The loads in the members that connected at U10W had increased over time, and the concentrated construction materials and equipment led to a significant increase in load, but the loads at the time of the collapse were still far below the level that should have been necessary to cause failure. The FHWA global finite element model showed that the L9/U10W diagonal was only about 5 percent above its design load, while the other members connected at the U10W node were below their design loads. The design load is used to determine the cross-sectional area and moment of inertia of the members to meet the allowable stress design criteria. Under the design load, the axial stress in each member must be less than the applicable allowable stress, which would be a maximum of 55 percent of the yield stress. Thus, even if placed under a load a few percent above its design load, each member would be expected to retain a sizable additional load capacity.

The original design of the bridge had additional conservatism built in. The design load includes the dead load from the weight of the structure plus a live load and an impact load. The live load and impact load components were calculated using AASHO-specified loads to simulate rows of heavy trucks across the bridge. The live load for each member was to be that combination of AASHO-specified loads that had the largest effect for that member. The impact load was calculated by multiplying the live load by a factor depending on span length (9 percent in the center span). The live load and impact load components were therefore intended to allow for a large, but unusual, load case. For L9/U10W, the dead loads from the modifications and the traffic and construction loads at the time of the collapse were slightly greater than the allowance for live load included in the original design, so the load in this member was slightly above its design load. However, the allowable stress design methodology that was correctly used to design the members ensured that member L9/U10 had more than adequate capacity to safely carry this load.

The U10 gusset plates should have been designed to meet the same allowable stress requirements as the truss members. Because stresses arise in the gusset plates from balancing the loads in the truss members at each node, the design loads for the members should have been used to determine the design of the gusset plates. AASHO specifications required that the gusset plates be of ample thickness to resist shear, direct stress, and flexure, but the FHWA analysis showed that the

U10 gusset plates had inadequate capacity to meet those requirements under the member design loads. Had the U10 gusset plates been designed in accordance with AASHO specifications, they also would have had a sizable additional load capacity under the loads carried by the main truss members at the time of the collapse.

The Safety Board therefore concludes that the loading conditions that caused the failure of the improperly designed gusset plates at the U10 nodes included substantial increases in the dead load from bridge modifications and, on the day of the accident, the traffic load and the concentrated loads from the construction materials and equipment; if the gusset plates had been designed in accordance with AASHO specifications, they would have been able to safely sustain these loads, and the accident would not have occurred.

Guidance for Allowing Construction Loads on Bridges

Mn/DOT specifications required that the low-slump concrete used for the roadway overlay be mixed on site. The quick set-up time for the materials and time limits for pouring and screeding that were built into the specifications argued for mixing the concrete as close to the pour site as possible. On August 1, 2007, the pour was to extend from the center of the bridge (node 14 of the center deck truss span) to the north end of the deck truss. For the sake of efficiency, personnel from PCI (the contractor adding the overlay on the day of the accident) decided to stage the construction aggregates and equipment near the south end of the center span, just over node 10. Although Mn/DOT had no policy that specifically required contractors to obtain approval before stockpiling materials on a bridge, contractor employees indicated that on a previous occasion they had asked a Mn/DOT construction inspector about such stockpiling and had been given a response that they interpreted as permission.

According to PCI, the previous request for approval of stockpiling of materials on the deck reflected a concern about the time and effort that would be involved in moving the materials to another location if Mn/DOT were to later determine that they should be positioned elsewhere. No evidence was found that contractor personnel were concerned about the weight of the materials. Nor would the weight itself have been an obvious consideration in that the weight of the aggregates that were added to the deck truss portion of the bridge was less than the weight of the concrete that had been previously milled off.

When the aggregates were delivered, they occupied a space about 101 feet long and 20–26 feet wide. To facilitate traffic and construction work, they were later rearranged into an area 115 feet long and 12–16 feet wide. If the overlay project had been completed as planned, much of the weight of the aggregates (and the cement that would have been added to them) would have been spread over an area 530 feet long and 24 feet wide, and the undersized gusset plates would not have been overloaded. The problem with the construction loads was thus not the weight of the aggregates but the concentrated nature of the loading of the aggregates and associated construction equipment.

Mn/DOT officials told the Safety Board after the accident that they would probably have denied a request to stockpile materials as they were on the day of the collapse because of the concentrated loads. But because the contractor was not required to, and did not, formally ask for permission, it is impossible to know with certainty what the Mn/DOT response would have been. If PCI had made a written request to the project engineer regarding the stockpiling, one option for Mn/DOT would have been to base its decision on the result of load analysis, which would have indicated whether the proposed loads exceeded the allowable load limit of the structure. At the Safety Board's request, Mn/DOT performed such an analysis, which—though not addressing the capacity of the gusset plates—indicated that the structure should have been able to safely support the additional load. Had Mn/DOT made a decision based solely on such an analysis, it likely would have approved the stockpiling.

Although Mn/DOT officials stated that they would probably not have approved the concentration of loads as they were on the day of the collapse, there is no formal guidance that would have led them to that decision. AASHTO guidance on construction loading advised only that such loads should not exceed the load-carrying capacity of the structure. The technical advisory issued by the FHWA a week after the accident suggested that bridge owners should "ensure that any construction loading and stockpiled raw materials placed on a structure do not overload its members." Neither the advisory nor the existing guidance suggested how such an assurance was to be achieved.

A Safety Board survey of 10 State DOTs revealed that almost all rely heavily on the contractor for determining the safe placement of construction loads, and almost all are primarily concerned with oversized vehicles rather than with the potential stockpiling of raw materials. Of the 39 States that responded to an AASHTO survey, only 22 reported having procedures in place for the review of construction loads, including loads from stockpiled materials and construction equipment. Of these 22 States, the majority stated that the operation and storage of equipment/materials would be required to follow the State truck size and weight statutes.

In the absence of formal and specific guidance, decisions about the placement of construction materials may be made on an ad hoc basis or may be considered in the same way as an overweight vehicle permit, and may not take into account all the considerations necessary to ensure that temporary loads do not damage the structure or possibly even exceed the load-carrying capacity of the structure at its most highly stressed location.

The Safety Board concludes that without clear specifications and guidelines to direct bridge owners regarding the stockpiling of raw materials, they may fail to conduct the appropriate engineering reviews or analyses before permitting raw materials to be stockpiled on a bridge. The Safety Board believes that AASHTO should develop specifications and guidelines for use by bridge owners to ensure that construction loads and stockpiled raw materials placed on a structure during construction or maintenance projects do not overload the structural members or their connections.

During any bridge repaving or repair project, factors such as lane closures, tight construction schedules, and the necessity to navigate around previously repaired areas could contribute to the perceived need to stage aggregates on the bridge deck. Any guidance developed by AASHTO should direct bridge owners in addressing these issues, which could be clarified during the preconstruction conference. AASHTO could also include guidelines for improving the coordination between transportation offices and contractors to ensure clear definition of the approval process and expectations for each party. The apparent overreliance on contractors to exercise judgment about bridge loading when they are typically not qualified to do so should also be addressed. Finally, of course, these specifications and guidelines should require a sufficiently detailed engineering review and assessment to ensure that all affected members of a structure can support higher-than-normal loading.

The Safety Board notes that, following the collapse, Mn/DOT revised its *Standard Specifications for Construction* to address the storage of construction materials on bridges. Mn/DOT now limits such loads to a level commensurate with the normal design load of the structure unless the additional loading is part of the original design plans or is approved by the State bridge engineer. The revised Mn/DOT policy also addresses the manner of distributing such loads across the structure to prevent their concentration within a small area.

Inspections of I-35W Bridge

The I-35W bridge was subjected to four types of inspections. The first three inspections were requirements set by the National Bridge Inspection Standards and included routine inspections at intervals not to exceed 24 months, fracture-critical member inspections at intervals not to exceed 24 months, and underwater inspections at intervals not to exceed 60 months. Beginning in 1971, routine inspections were conducted annually on the I-35W bridge; and beginning in 1994, fracture-critical member inspections were conducted annually. Beginning in 2000, underwater inspections were conducted every 48 months. Thus, in each of these three areas, the I-35W bridge was subjected to more frequent inspections than required by the National Bridge Inspection Standards.

In addition to the requirements of the National Bridge Inspection Standards, Mn/DOT had initiated inspections of the I-35W bridge because of fatigue cracking that had been found in other Minnesota bridges similar in age. Mn/DOT had contracted with the University of Minnesota and URS to perform these inspections and analyses of the bridge and to assess the likelihood that fracture-critical support members could fail as a result of fatigue cracking. Based on the most recent URS evaluation in 2006, Mn/DOT initially allocated $1.5 million to retrofit (by adding reinforcing plates) each of the 52 identified fracture-critical structural members to prevent failure due to fatigue cracking. The retrofit project was to begin in January 2008. When nondestructive evaluation showed potential as an effective alternative to the retrofit, Mn/DOT delayed beginning the retrofit project until the effectiveness of these alternative methods could be evaluated. This evaluation was underway when the bridge collapsed.

In addition to inspection frequency, the National Bridge Inspection Standards also provide procedures regarding inspection criteria. Mn/DOT inspectors used these criteria as they inspected the bridge, including the gusset plates. Gusset plate conditions addressed by the criteria included corrosion, paint cracking and peeling, missing rivets, or other deficiencies with the fasteners. Although these inspections were capable of quantifying the condition of the gusset plates, they were not intended to analyze the adequacy of the gusset plate design. The Safety Board concludes that, although the I-35W bridge had been inspected in accordance with the National Bridge Inspection Standards and more frequently than required by the standards, these inspections would not have been expected to detect design errors. The additional inspections and analyses arranged for by Mn/DOT did indicate, however, that the agency was taking steps to address known risks to the bridge.

Condition Rating

Based on the results of its routine inspections, the I-35W bridge had been rated *Structurally Deficient* by the FHWA since 1991, when the superstructure received its first condition rating of 4 (poor condition). The bridge superstructure continued to have a recorded condition rating of 4 on each of the National Bridge Inventory forms between 1991 and 2007, including 1999 when the condition rating was not properly submitted to the FHWA.

The I-35W bridge was one of more than 72,000 bridges in the National Bridge Inventory that were rated as *Structurally Deficient*, and one of 145 (of a total of 465) steel deck truss bridges so rated. Thus, a status of *Structurally Deficient* is not uncommon, and such bridges are located in every State and U.S. territory. As of December 2007, no State or territory had fewer than 20 bridges rated as *Structurally Deficient*.

None of the conditions that were rated "poor" on the I-35W bridge superstructure involved gusset plates. Inspectors noted such conditions as failing paint, surface rust and corrosion, and some section loss on almost every type of main truss structural member. Inspectors also noted numerous poor weld details on the main truss members, and some bearing assemblies appeared to have insufficient movement. None of these conditions were considered to be a threat to the load-bearing capacity of the structure, and none were determined by this investigation to have played a role in the collapse. The Safety Board therefore concludes that, although the I-35W bridge had been rated under the National Bridge Inspection Standards as *Structurally Deficient* for 16 years before the accident, the conditions responsible for that rating did not cause or contribute to the collapse of the bridge.

Gusset Plate Corrosion

The Safety Board reviewed the reports from both routine and fracture-critical inspections of the I-35W bridge, giving particular attention to the condition of gusset plates. The first specific reference to gusset plates occurred in the 1993 routine inspection report as corrosion having been found on at least one gusset plate at the L11E node. Subsequent routine inspection reports from 1994–2006 did not specifically note corrosion on this or any other gusset plate on the bridge, though each routine report had general comments about pack rust and section loss on unspecified structural members.

Each of the fracture-critical inspection reports from 1994–2006 did note the presence of rust, corrosion, and section loss on gusset plates, but in most cases, the reports simply repeated condition comments from previous inspections. Inspectors did not quantify the amount of rust or corrosion and apparently had not attempted to thoroughly assess the degree or rate of deterioration.

When Safety Board investigators examined the gusset plates at the L11E and L11W nodes as part of this accident investigation, they found corrosion and loss of section on all four of the gusset plates. The 1993 routine inspection report cited an 18-inch-long line of corrosion on the inside (west) gusset plate at the L11E node (the only plate specifically identified in a Mn/DOT bridge inspection report). Postcollapse examination found corrosion over almost the entire 99-inch length of this gusset plate as well as on the other gusset plates from the L11 nodes. This significant increase in the area of corrosion and resulting section loss had not been noted on the many routine inspection reports between 1993 and 2006, though the presence of corrosion on this gusset plate had been noted in subsequent fracture-critical inspections. Postcollapse measurements of the gusset plates at the L11E and L11W nodes revealed an average total section loss due to corrosion ranging from 4.7 percent for the outside (east) gusset plate at L11E to 17.1 percent for the inside (west) gusset plate at L11E. As a measure of the significance of the corrosion, the rivet holes immediately adjacent to the line of corrosion accounted for a section loss of about 27 percent.

Although the investigation of this accident showed that the corrosion found on the L11 gusset plates played no role in the collapse of the I-35W bridge, the Safety Board is concerned that the amount of corrosion and section loss that was observed on the L11 gusset plates had not been documented in detail and had not been given particular attention during subsequent bridge inspections. Had the bridge remained in service, and had the corrosion and section loss continued to progress without mitigation, the ability of the gusset plates to safely carry loads would have continued to diminish, even if these particular plates had been properly sized.

Gusset Plate Bowing

Before the collapse, seven of the eight U10 and U10' gusset plates on the I-35W bridge had noticeable distortion in the form of bowing along one unsupported edge of each plate. This distortion was visible in photographs taken in 1999 and 2003, following both the 1977 increase in deck thickness and the 1998 modification to the median barrier and outside traffic railings. The Safety Board found no pre-1998 photographs of the gusset plates. Distortion of the plate edges was visible only between compression diagonal L9(')/U10(') and upper chord member U9(')/U10('). The U10E and U10W gusset plates were bowed to the west, and three of the four U10'E and U10'W gusset plates were bowed to the east. (The photographs were insufficient to establish the presence of bowing in the outside gusset plate at the U10' node.) The measured magnitude of bowing on the inside plates at the four U10(') nodes ranged from 0.44–0.99 inch.

The ratio between the length of the unsupported edges of these gusset plates (about 30 inches) and their thickness (specified as 0.5 inch) was 60, which exceeded the ratio of 48, above which AASHO design specifications required the edge to be stiffened. The bowing was reportedly observed by a Mn/DOT Metro District bridge safety inspection engineer, but it was believed to have occurred during construction of the bridge, so no note of bowing was made on any inspection report, and no evidence was found that Mn/DOT did any analysis to determine why the distortion had occurred or whether or to what extent it had affected the load-carrying capacity of the gusset plates.

The Safety Board considered what might have caused the bowed gusset plates at the U10 and U10' nodes. Possible sources of this distortion include loads placed on the members during erection (either before or after closure of the structure at the center of the center span), changes in geometry during erection, unusual loads generated as the members were connected during erection, large dead loads experienced during the life of the bridge, or repetitive large live loads experienced during the operational life of the bridge. Some combination of all of these types of loads could also have contributed to bowing of the gusset plates.

The loads generated in the members of the main truss at the U10(') nodes during the steel erection process were evaluated and found to be much lower than the loads in these members after construction was complete and the deck was poured. Also, changes in the geometry of the truss from the preclosed condition to the fully constructed condition were determined to be minimal. The normal loads that would be encountered during erection of the bridge were therefore insufficient to cause the bowing of any of these gusset plates. The Safety Board was unable to fully evaluate the possibility that some type of unusual loading was generated in the gusset plates as the L9(')/U10(') diagonal was connected to the U9(')/U10(') upper chord member but considers it unlikely that such loading could have caused bowing deformation at all four U10(') nodes.

With regard to a sustained dead load, the largest dead load before August 1, 2007, likely occurred during the 1998 modifications, when rows of temporary barriers were placed on the bridge as the median and outside barriers were increased in size. A finite element analysis of this load case was performed to investigate whether this scenario could have caused bowing of the gusset plates. The finite element analysis showed that the loads in the L9/U10 diagonals were near their design load with the modified median barrier and outside traffic railings and the temporary barriers in place. Because the bowing would likely be sensitive to initial imperfections, an initial bowing magnitude of 0.05 inch was included in the model. However, at the end of the analysis, after the weight of the temporary barriers was removed, the final maximum bowing magnitude was only 0.11 inch, much less than the bowing recorded in the 1999 or 2003 photographs. Further, the finite element analysis also showed that an initial bowing magnitude of about 0.5 inch was necessary to have a bowing magnitude of 0.6 inch to match the photographs under the loading conditions when the photographs were taken. These results indicate that a single large load application after the bridge was built is also unlikely to have caused the bowing.

A third possible source of the bowing distortion is repetitive live loads over the course of the 40-year life of the bridge. The load case looking at the effects of the temporary barriers did show a small permanent increase in bowing magnitude following the application of a single large load. Repetitive applications of large live loads could therefore have had a cumulative effect and caused the bowing of the magnitude observed. However, for those gusset plates that were photographed in both 1999 and 2003, the measurements of bowing magnitude from 2003 were not significantly different from those shown in the 1999 photograph.

Because of uncertainties related to any of the mechanisms that were considered as possibly having caused the bowing distortion, the Safety Board was unable to determine precisely when and how this distortion was generated. Regardless of the source of the bowing, however, properly designed U10(') node gusset plates would have had significantly greater load capacity and would have been thick enough to resist the forces that caused the bowing. The Safety Board concludes that the bowing of the gusset plates at the U10(') nodes was symptomatic of the inadequate capacity of the plates and occurred under an undetermined load condition before 1999.

Inspecting U.S. Bridges for Gusset Plate Adequacy

Corrosion on Gusset Plates

The I-35W bridge was only one of a number of steel truss bridges that were found to have gusset plate corrosion and section loss that had been overlooked or underestimated by State bridge inspectors. In 1996, gusset plates on the eastbound Lake County Grand River bridge in Ohio failed while the bridge was undergoing maintenance. The failure was attributed to corrosion and section loss, which had completely penetrated the gusset plates at some locations. The amount of section loss had been masked by corrosion products to the extent that it could not be adequately assessed solely through visual bridge inspections.

Similarly, ODOT discovered in October 2007 that visual inspections of the gusset plates on the Cuyahoga County Innerbelt bridge in Cleveland had significantly underestimated the amount of section loss. The actual degree of section loss in the gusset plates was determined only through the use of nondestructive evaluation methods, specifically hand-held ultrasonic thickness gauges.

More recently, in the wake of the collapse of the I-35W bridge, Mn/DOT conducted detailed inspections and analyses of 25 other truss bridges in the State and found significant corrosion and section loss on the Highway 43 bridge in Winona, Minnesota. The amount of section loss in some of the plates was sufficient to prompt Mn/DOT to close the bridge until an analysis could be performed to determine the safe capacity of the bridge in light of the deteriorated gusset plates. A fracture-critical inspection had been completed on this bridge on August 1, 2007, the day the I-35W bridge collapsed. The report of this inspection noted severe deterioration in some of the gusset plates but nonetheless concluded that the bridge had no critical structural deficiencies. The report recommended that some cracked welds in the bottom chord of the deck truss be monitored during future inspections, but it made no recommendation for more frequent or in-depth inspection or monitoring of the deteriorated gusset plates.

A routine inspection of the bridge in Winona had been conducted less than 4 months before the fracture-critical inspection, but it too had identified no critical findings with regard to the bridge superstructure. The report did note the presence of rust between some of the gusset plates and their steel members but concluded that "the connections are still functioning." Although this statement was accurate in that the bridge had not fallen, the Safety Board is concerned that inspectors did not attempt either to address the reduction in load-carrying capacity that might have resulted from the existing section loss or to assess the potential effects of any further deterioration of the gusset plates.

The detection of corrosion in gusset connections is often hampered by the configuration of the connection. The insides of gusset plates, which are perhaps the most susceptible to corrosion, are often difficult to inspect visually even if a concentrated effort is made beforehand to clean or remove debris from the connection. Further, when corrosion is found, its surface appearance often belies the actual amount of section loss that has occurred. Thus, State DOTs whose inspectors rely solely on visual examination to quantify the amount of corrosion on gusset plates or to assess its potential to weaken the connection do not have sufficient information to make accurate quantifications or assessments. The Safety Board therefore concludes that because visual bridge inspections alone, regardless of their frequency, are inadequate to always detect corrosion on gusset plates or to accurately assess the extent or progression of that corrosion, inspectors should employ appropriate nondestructive evaluation technologies when evaluating gusset plates.

The Safety Board believes that the FHWA should require that bridge owners assess the truss bridges in their inventories to identify locations where visual inspections may not detect gusset plate corrosion and where, therefore, appropriate nondestructive evaluation technologies should be used to assess gusset plate condition.

The Safety Board commends the FHWA for its "Bridge Inspector's NDE Showcase," the 1-day training program developed to demonstrate commercially available advanced bridge evaluation and inspection tools. The expanded use of these existing technologies to supplement visual inspections should result in more precise evaluations of bridge components.

Distortion of Gusset Plates

In its review of all State steel truss bridges after the I-35W bridge collapse, Mn/DOT also detected distortion in the gusset plates of the DeSoto bridge in St. Cloud, Minnesota. The consultant retained to evaluate and determine the origin of the distortion found that it likely had occurred when the bridge was built in 1957. Thus, though this bridge had existed for 50 years, the distortion in the gusset plates was discovered only by an inspection that focused on the condition of gusset plates, an emphasis that had not been evident in previous routine or fracture-critical inspections. Also, like the I-35W bridge gusset plates, the distorted gusset plates on the DeSoto bridge did not meet the standards for the length of unsupported edge, again suggesting that the plates had not been subjected to adequate design review. Because of the gusset plate condition, the DeSoto bridge was closed in March 2008, and its replacement date was moved up from 2015 to 2008.

About 2 months after the I-35W bridge collapse, Ohio inspectors, in addition to finding significant corrosion on the gusset plates of the Cuyahoga County Innerbelt bridge, found that some of the corroded plates were also distorted. It is unlikely that this distortion would have been noted except for the increased emphasis on verifying the integrity of steel truss bridge superstructures, including gusset plates. The Safety Board concludes that distortion such as bowing is a sign of an out-of-design condition that should be identified and subjected to further engineering analysis to ensure that the appropriate level of safety is maintained.

Guidance for Inspecting Gusset Plates

The majority of States base their bridge inspections on the Pontis bridge management software program. But Pontis, like other bridge management systems, does not include gusset plates because the *AASHTO Guide for Commonly Recognized (CoRe) Structural Elements* does not include gusset plates as a bridge structural element requiring specific attention and subsequent condition rating during bridge inspections. Gusset plates can be noted using the "smart flag" system, but no specific action is required if a gusset plate condition is so noted. By not including gusset plates as separate inspection elements with specific condition rating guidelines, the AASHTO guidance (and the bridge management systems that are based on it) may lead bridge owners and inspectors to give inadequate attention to these critical bridge components. The Safety Board concludes that because the *AASHTO Guide for Commonly Recognized (CoRe) Structural Elements* does not include gusset plates as a separate bridge inspection element, bridge owners may fail to adequately document and track gusset plate conditions that could threaten the safety of the structure. The Safety Board believes that AASHTO should include gusset plates as a CoRe structural element and develop guidance for bridge owners in tracking and responding to potentially damaging conditions in gusset plates, such as corrosion and distortion; and revise the *AASHTO Guide for Commonly Recognized (CoRe) Structural Elements* to incorporate this new information.

Training for Bridge Inspectors

The primary guidance document for bridge inspectors is the FHWA *Bridge Inspector's Reference Manual*. A number of topics and subtopics in the manual are applicable to the inspection of steel truss bridges. Although these sections address types of steel and steel deterioration, steel failure mechanics, and procedures and locations for inspecting bridge structural members, including fracture-critical members, none of them specifically refer to gusset plates on main truss members.

The minimum qualifications for bridge inspectors, including the requirements for experience and training, are spelled out in the National Bridge Inspection Standards. All bridge inspection project managers and team leaders must complete a bridge inspection training program approved by the FHWA. The National Highway Institute provides a 3-week training program for bridge inspectors that consists of a 1-week "Engineering Concepts for Bridge Inspectors" course and a 2-week "Safety Inspection of In-Service Bridges" course. When combined, these courses, which are based on the *Bridge Inspector's Reference Manual*, meet the requirements for a comprehensive training program as defined in the National Bridge Inspection Standards. The National Highway Institute also offers a 3.5-day course for bridge inspectors, "Fracture Critical Inspection Techniques for Steel Bridges."

Safety Board investigators reviewed the training materials for these courses and found only a few, very general, references to gusset plates. None of the materials emphasized the importance of gusset plates as structural members or identified deficiencies, such as distortion, that should be of particular concern to inspectors. This lack of emphasis on gusset plates in the FHWA primary bridge inspection reference document and in the National Highway Institute training courses could, in part, explain the apparent lack of due attention to gusset plate condition exhibited by bridge inspectors in several States and identified during this accident investigation. The Safety Board concludes that the lack of specific references to gusset plates in the *Bridge Inspector's Reference Manual* and in National Highway Institute bridge inspector training courses could cause State bridge inspectors during routine or fracture-critical bridge inspections to fail to give appropriate attention to distortions, such as bowing, in gusset plates.

The Safety Board believes that the FHWA should modify the approved bridge inspector training as follows: (1) update the National Highway Institute training courses to address inspection techniques and conditions specific to gusset plates, emphasizing issues associated with gusset plate distortion as well as the use of nondestructive evaluation at locations where visual inspections may be inadequate to assess and quantify such conditions as section loss due to corrosion; and, (2) at a minimum, include revisions to reference material, such as the *Bridge Inspector's Reference Manual*, and address any newly developed gusset plate condition ratings in the AASHTO commonly recognized (CoRe) structural elements.

CONCLUSIONS

Findings

1. The initiating event in the collapse of the I-35W bridge was a lateral shifting instability of the upper end of the L9/U10W diagonal member and the subsequent failure of the U10 node gusset plates on the center portion of the deck truss.

2. Because the deck truss portion of the I-35W bridge was non-load-path-redundant, the total collapse of the deck truss was likely once the gusset plates at the U10 nodes failed.

3. The examination of the collapsed structure, the finite element analysis, and the video recording of the collapse showed that the following were neither causal nor contributory to the collapse of the I-35W bridge: corrosion damage found on the gusset plates at the L11 nodes and elsewhere, fracture of a floor truss, preexisting cracking in the bridge deck truss or approach spans, temperature effects, or shifting of the piers.

4. The initial emergency response to the bridge collapse by fire and rescue units was timely and appropriate, and the incident command system was well coordinated.

5. The damage to bridge components that occurred during victim recovery did not, in this case, prevent determination of the collapse sequence.

6. The gusset plates at the U10 nodes, where the collapse initiated, had inadequate capacity for the expected loads on the structure, even in the original as-designed condition.

7. Because the bridge's main truss gusset plates had been fabricated and installed as the designers specified, the inadequate capacity of the U10 node gusset plates had to have been the result of an error on the part of the bridge design firm.

8. Even though the bridge design firm knew how to correctly calculate the effects of stress in gusset plates, it failed to perform all necessary calculations for the main truss gusset plates of the I-35W bridge, resulting in some of the gusset plates having inadequate capacity, most significantly at the U4 and U4', U10 and U10', and L11 and L11' nodes.

9. Although the U10 gusset plates would have required edge stiffeners according to American Association of State Highway Officials specifications, the addition of stiffeners would not have made the U10 gusset plates adequate or prevented the gusset plates from yielding.

10. The design review process used by the bridge design firm was inadequate in that it did not detect and correct the error in design of the gusset plates at the U4 and U4', U10 and U10', and L11 and L11' nodes before the plans were made final.

11. Neither Federal nor State authorities evaluated the design of the gusset plates for the I-35W bridge in sufficient detail during the design and acceptance process to detect the design errors in the plates, nor was it standard practice for them to do so.

12. Current Federal and State design review procedures are inadequate to detect design errors in bridges.

13. Because current American Association of State Highway and Transportation Officials guidance directs bridge owners to rate their bridges when significant changes occur but not before they place new bridges in service, the load-carrying capacity of new bridges may not be verified before they are opened to traffic.

14. Had American Association of State Highway and Transportation Officials guidance included gusset plates in load ratings, there would have been multiple opportunities to detect the inadequate capacity of the U10 gusset plates of the I-35W bridge deck truss.

15. Because bridge owners generally consider gusset plates to be designed more conservatively than the other members of a truss, because the American Association of State Highway and Transportation Officials provides no specific guidance for the inspection of gusset plates, and because commonly used computer programs for load rating analysis do not include gusset plates, bridge owners typically ignore gusset plates when performing load ratings, and the resulting load ratings might not accurately reflect the actual capacity of the structure.

16. The loading conditions that caused the failure of the improperly designed gusset plates at the U10 nodes included substantial increases in the dead load from bridge modifications and, on the day of the accident, the traffic load and the concentrated loads from the construction materials and equipment; if the gusset plates had been designed in accordance with American Association of State Highway Officials specifications, they would have been able to safely sustain these loads, and the accident would not have occurred.

17. Without clear specifications and guidelines to direct bridge owners regarding the stockpiling of raw materials, they may fail to conduct the appropriate engineering reviews or analyses before permitting raw materials to be stockpiled on a bridge.

18. Although the I-35W bridge had been inspected in accordance with the National Bridge Inspection Standards and more frequently than required by the standards, these inspections would not have been expected to detect design errors.

19. Although the I-35W bridge had been rated under the National Bridge Inspection Standards as *Structurally Deficient* for 16 years before the accident, the conditions responsible for that rating did not cause or contribute to the collapse of the bridge.

20. The bowing of the gusset plates at the U10 and U10' nodes was symptomatic of the inadequate capacity of the plates and occurred under an undetermined load condition before 1999.

21. Because visual bridge inspections alone, regardless of their frequency, are inadequate to always detect corrosion on gusset plates or to accurately assess the extent or progression of that corrosion, inspectors should employ appropriate nondestructive evaluation technologies when evaluating gusset plates.

22. Distortion such as bowing is a sign of an out-of-design condition that should be identified and subjected to further engineering analysis to ensure that the appropriate level of safety is maintained.

23. Because the *AASHTO Guide for Commonly Recognized (CoRe) Structural Elements* does not include gusset plates as a separate bridge inspection element, bridge owners may fail to adequately document and track gusset plate conditions that could threaten the safety of the structure.

24. The lack of specific references to gusset plates in the *Bridge Inspector's Reference Manual* and in National Highway Institute bridge inspector training courses could cause State bridge inspectors during routine or fracture-critical bridge inspections to fail to give appropriate attention to distortions, such as bowing, in gusset plates.

Probable Cause

The National Transportation Safety Board determines that the probable cause of the collapse of the I-35W bridge in Minneapolis, Minnesota, was the inadequate load capacity, due to a design error by Sverdrup & Parcel and Associates, Inc., of the gusset plates at the U10 nodes, which failed under a combination of (1) substantial increases in the weight of the bridge, which resulted from previous bridge modifications, and (2) the traffic and concentrated construction loads on the bridge on the day of the collapse. Contributing to the design error was the failure of Sverdrup & Parcel's quality control procedures to ensure that the appropriate main truss gusset plate calculations were performed for the I-35W bridge and the inadequate design review by Federal and State transportation officials. Contributing to the accident was the generally accepted practice among Federal and State transportation officials of giving inadequate attention to gusset plates during inspections for conditions of distortion, such as bowing, and of excluding gusset plates in load rating analyses.

RECOMMENDATIONS

As a result of its investigation of the collapse of the I-35W bridge in Minneapolis, Minnesota, the National Transportation Safety Board makes the following safety recommendations:

New Recommendations

To the Federal Highway Administration:

> Develop and implement, in conjunction with the American Association of State Highway and Transportation Officials, a bridge design quality assurance/quality control program, to be used by the States and other bridge owners, that includes procedures to detect and correct bridge design errors before the design plans are made final; and, at a minimum, provides a means for verifying that the appropriate design calculations have been performed, that the calculations are accurate, and that the specifications for the load-carrying members are adequate with regard to the expected service loads of the structure. (H-08-17)

> Require that bridge owners assess the truss bridges in their inventories to identify locations where visual inspections may not detect gusset plate corrosion and where, therefore, appropriate nondestructive evaluation technologies should be used to assess gusset plate condition. (H-08-18)

> Modify the approved bridge inspector training as follows: (1) update the National Highway Institute training courses to address inspection techniques and conditions specific to gusset plates, emphasizing issues associated with gusset plate distortion as well as the use of nondestructive evaluation at locations where visual inspections may be inadequate to assess and quantify such conditions as section loss due to corrosion; and, (2) at a minimum, include revisions to reference material, such as the *Bridge Inspector's Reference Manual*, and address any newly developed gusset plate condition ratings in the American Association of State Highway and Transportation Officials commonly recognized (CoRe) structural elements. (H-08-19)

To the American Association of State Highway and Transportation Officials:

Work with the Federal Highway Administration to develop and implement a bridge design quality assurance/quality control program, to be used by the States and other bridge owners, that includes procedures to detect and correct bridge design errors before the design plans are made final; and, at a minimum, provides a means for verifying that the appropriate design calculations have been performed, that the calculations are accurate, and that the specifications for the load-carrying members are adequate with regard to the expected service loads of the structure. (H-08-20)

Revise your *Manual for Bridge Evaluation* to include guidance for conducting load ratings on new bridges before they are placed in service. (H-08-21)

Modify the guidance and procedures in your *Manual for Bridge Evaluation* to include evaluating the capacity of gusset plates as part of the load rating calculations performed for non-load-path-redundant steel truss bridges. (H-08-22)

When the findings of the Federal Highway Administration–American Association of State Highway and Transportation Officials joint study on gusset plates become available, update the *Manual for Bridge Evaluation* accordingly. (H-08-23)

Develop specifications and guidelines for use by bridge owners to ensure that construction loads and stockpiled raw materials placed on a structure during construction or maintenance projects do not overload the structural members or their connections. (H-08-24)

Include gusset plates as a commonly recognized (CoRe) structural element and develop guidance for bridge owners in tracking and responding to potentially damaging conditions in gusset plates, such as corrosion and distortion; and revise the *AASHTO Guide for Commonly Recognized (CoRe) Structural Elements* to incorporate this new information. (H-08-25)

Previously Issued Recommendation Resulting From This Accident Investigation

As result of its investigation of this accident, the Safety Board issued the following safety recommendation to the Federal Highway Administration on January 15, 2008:

> For all non-load-path-redundant steel truss bridges within the National Bridge Inventory, require that bridge owners conduct load capacity calculations to verify that the stress levels in all structural elements, including gusset plates, remain within applicable requirements whenever planned modifications or operational changes may significantly increase stresses. (H-08-1)

BY THE NATIONAL TRANSPORTATION SAFETY BOARD

MARK V. ROSENKER
Acting Chairman

DEBORAH A. P. HERSMAN
Member

KATHRYN O'LEARY HIGGINS
Member

ROBERT L. SUMWALT
Member

STEVEN R. CHEALANDER
Member

Adopted: November 14, 2008

APPENDIX A

Investigation

The National Transportation Safety Board was notified of the Minneapolis, Minnesota, bridge collapse on August 1, 2007. Investigative teams were dispatched from the Safety Board's Washington, D.C.; Atlanta, Georgia; Arlington, Texas; and Gardena, California, offices. Separate groups were established to investigate structural engineering, bridge design, construction oversight, and survival factors issues. Other groups were formed to facilitate evidence documentation, structural modeling, and witness identification. Chairman Mark Rosenker was the Board Member on scene.

Participating in the on-scene investigation were representatives of the Federal Highway Administration, the Minnesota Department of Transportation, the Minnesota State Police, the Minneapolis Police Department, the Hennepin County Sheriff's Office, and the maintenance contractor, Progressive Contractors, Inc. Jacobs Engineering (the company that had acquired the firm responsible for original design of the bridge) initially provided design plans and other related documents and later, on January 17, 2008, was included as an official party to the investigation.

The on-scene investigation, including documentation and analysis of the recovered bridge structure, required Safety Board investigators and other support staff to remain at the accident site from August 2–November 10, 2007.

APPENDIX B

Selected Bridge Information

Table B-1. Basic bridge information from selected States.

State DOT	Total districts/ regions	Total State bridges^A/local bridges	% Consultant % In-house bridge designs	Bridge load rating programs	Bridge management system
California	12 districts	12,185 – State 11,782 – local	50% consultant 50% in-house	Virtis	Pontis with modifications
Florida	8 districts	6,068 – State 5,532 – local	95% consultant 5% in-house	See note 1	Pontis with modifications
Iowa	6 districts	4,064 – State 20,360 – local	60% consultant 40% in-house (FY2008)	See note 2	See note 3
Kansas	6 districts	4,940 – State 20,524 – local	70% consultant 30% in-house	See note 4	Pontis with modifications
Maryland	7 districts	2,578 – State 2,233 – local	50% consultant 50% in-house	See note 5	Inventory and appraisal information entered and stored in access database
Minnesota	8 districts (including Metro)	3,585 – State 9,344 – local	50% consultant 50% in-house	See note 6	Pontis with modifications
Nebraska	8 districts	3,511 – State 11,828 – local	**Statewide** 5% - consultant 95% in-house **Local** 95% consultant 5% in-house	See note 7	Pontis and in-house programs
New York	11 regions	7,632 – State 9,682 – local	50% consultant 50% in-house	See note 8	Pontis and in-house analysis tools
Ohio	12 districts	11,103 – State 17,974- local	95% consultant 5% in-house	PC BARS	Database monitored monthly using data mining software and spreadsheets
Oregon	5 regions	2,672 – State 3,974 – local	**Current** 20% consultant 80% in-house **Goal** 70% consultant 30% in-house	BRASS	Pontis with modifications

State DOT	Total districts/ regions	Total State bridges[A]/local bridges	% Consultant % In-house bridge designs	Bridge load rating programs	Bridge management system
Pennsylvania	11 districts	15,877 – State 6,416 – local	**Statewide** 60% consultant 40% in-house **Urban districts** 95% consultant 5% in-house	See note 9	Pontis with modifications
Tennessee	4 regions	8,150 – State 11,419 – local	5% consultant 95% in-house	See note 10	Pontis and in-house analysis tools
Texas	25 districts	33,028 – State 17,448 – local	40% consultant 60% in-house	See note 11	Pontis with modifications
Virginia	9 districts	11,721 – State 1,416 – local	30% consultant 70% in-house	See note 12	Pontis and HTRIS
Washington	7 regions	3,019 – State 3,878 – local	10% consultant 90% in-house	BRIDG FOR WINDOWS	Pontis with modifications

[A] The table shows bridges or culverts that carry vehicular traffic and are longer than 20 feet as defined by the National Bridge Inventory. Bridges on a toll authority system are included in the total number of local bridges.

Table B-1 Notes:

1 Florida bridge load rating programs include Leap Conspan, Smart Bridge, STAAD, BRUFEM, Merlin-Dash, GT STRUDEL, BAR7, MIDAS, BDAC, MDX, ADAPT, PC BARS, Virtis, and Smartbridge.

2 Iowa bridge load rating programs include LARS and Virtis.

3 The Iowa DOT is in the process of implementing PONTIS as an additional tool to help identify candidates for the Transportation Improvement Program.

4 Kansas bridge load rating programs include Virtis, STAAD, and BRASS.

5 Maryland bridge load rating programs include Merlin-Dash, BARS5, BARS7, STAAD, and in-house spreadsheets.

6 Minnesota bridge load rating programs include BARS and Virtis.

7 Nebraska bridge load rating programs include BARS, LARS, Virtis, and in-house programs.

8 New York State bridge load rating programs include Virtis and BLRS (Bridge Load Rating System).

9 Pennsylvania bridge load rating programs include BAR7, STAAD, and BSDI-3D.

10 Tennessee bridge load rating programs include Virtis, BARS, Conspan, and Excel spreadsheets.

11 Texas bridge load rating programs include BMCOL51, PSTRS14, RISA, STAAD, BRASS, and RATE.

12 Virginia bridge load rating programs include BARS, Virtis, DESCUS (Curved Girder Program), and STAAD.

Table B-2. Total number of bridges by deck area in selected States.

State DOT	Total number of State/ local bridges	Total deck area (square feet)
California	12,185 – State 11,782 - local	237,998,721 – State 64,494,529 – local
Florida	6,068 – State 5,532 – local	125,431,994 – State 38,257,764 – local
Iowa	4,064 – State 20,360 – local	35,434,725 – State 40,069,720 – local
Kansas	4,940 – State 20,524 – local	38,791,815 – State 45,469,949 – local
Maryland	2,578 – State 2,233 – local	28,441,714 – State 21,360,354 – local
Minnesota	3,585 – State 9,344 – local	47,027,471 – State 28,272,722 – local
Nebraska	3,511 – State 11,828 – local	22,090,847 – State 18,778,942 – local
New York	7,632 – State 9,682 – local	78,622,000 – State 57,345,000 – local
Ohio	11,103 – State 17,974- local	106,739,000 – State 34,778,100 – local
Oregon	2,672 – State 3,974 - local	35,125,249 – State 13,688,325 – local
Pennsylvania	15,877 – State 6,416 – local	106,503,300 – State 14,206,400 – local
Tennessee	8,150 – State 11,419 – local	78,203,975 – State 26,332,721 – local
Texas	33,028 – State 17,448 - local	366,973,079 – State 71,614,950 – local
Virginia	11,721 – State 1,416 – local	83,390,530 – State 20,051,368 – local
Washington	3,019 – State 3,878 – local	45,567,272 – State 14,187,731 – local

Appendix C

Previous Safety Board Actions Regarding Bridge Inspections

The Safety Board has a history of investigating bridge accidents, beginning with its investigation of the December 15, 1967, collapse of the Silver bridge over the Ohio River, in Point Pleasant, West Virginia, which killed 46 people.[1] The Safety Board determined that the probable cause of the bridge collapse was a fracture of an eyebar. The fracture had developed over the 40-year life of the structure due to a combination of stress corrosion and corrosion fatigue. The safety recommendations prompted the Federal Highway Administration (FHWA) to establish inspection standards for locating, inspecting, evaluating, and correcting bridge deficiencies. These standards eventually led to the establishment by Congress of the Highway Bridge Replacement and Rehabilitation Program, the precursor of today's bridge inspection programs.

On June 28, 1983, in Greenwich, Connecticut, a 100-foot-long suspended span of the Interstate 95 highway bridge over the Mianus River collapsed and fell 70 feet into the river.[2] The collapse resulted in three fatalities to vehicle occupants. The Safety Board determined that the probable cause of the collapse was the undetected lateral displacement, caused by corrosion-induced forces, of the hangers of the pin-and-hanger suspension assembly. This deficiency had gone undetected by routine inspections. As a result of this investigation, Safety Board recommendations were issued that led to development of the FHWA's fracture-critical bridge inspection program.

Prior to 1985, the FHWA had not emphasized underwater inspections or required its division offices to review a State's underwater inspection capabilities. On April 24, 1985, the U.S. 43 Chickasabogue bridge, near Mobile, Alabama,[3] collapsed due to scour.[4] As a result of the Safety Board's investigation of the collapse, the FHWA, in June 1985, addressed multiple issues leading to an increased emphasis on underwater inspection programs. The Safety Board also investigated

---

[1] National Transportation Safety Board, *Collapse of U.S. Highway Bridge, Point Pleasant, West Virginia, December 15, 1967*, Highway Accident Report NTSB/HAR-71/01 (Washington, DC: NTSB, 1971).

[2] National Transportation Safety Board, *Collapse of a Suspended Span of Interstate Route 95 Highway Bridge Over the Mianus River, Greenwich, Connecticut, June 28, 1983*, Highway Accident Report NTSB/HAR-84/03 (Washington, DC: NTSB, 1984).

[3] National Transportation Safety Board, *Collapse of the U.S. 43 Chickasabogue Bridge Spans Near Mobile, Alabama, April 24, 1985*, Highway Accident Report NTSB/HAR-86/01 (Washington, DC: NTSB, 1986).

[4] *Scour* refers to the erosion or removal of streambed or bank material from bridge foundations due to flowing water. Scour is the most common cause of highway bridge failures in the United States.

the April 1987 collapse of the Schoharie Creek bridge in Amsterdam, New York,[5] which resulted in 10 fatalities; and the April 1989 collapse of the Hatchie River bridge in Covington, Tennessee, which resulted in 8 fatalities.[6] Recommendations from these investigations led to the development of a comprehensive underwater inspection program.

In total, these five accidents resulted in 67 fatalities. As a result of its investigations of these accidents, the Safety Board issued safety recommendations that resulted in not only the development of a National Bridge Inspection Program but also significant improvements to that program, such as the institution of fracture-critical and underwater inspections.

[5] National Transportation Safety Board, Collapse of New York Thruway (I-90) Bridge Over the Schoharie Creek, Near Amsterdam, New York, April 5, 1987, Highway Accident Report NTSB/HAR-88/02 (Washington, DC: NTSB, 1988).

[6] National Transportation Safety Board, Collapse of the Northbound U.S. Route 51 Bridge Spans Over the Hatchie River Near Covington, Tennessee, April 1, 1989, Highway Accident Report NTSB/HAR-90/01 (Washington, DC: NTSB, 1990).

www.ingramcontent.com/pod-product-compliance
Lightning Source LLC
Chambersburg PA
CBHW080247180526
45167CB00006B/2442